76th Annual Meeting of the Labor and Employment Relations Association and LERA@ASSA Meeting 2024

Labor and Employment Relations Association Series

San Antonio, Texas, USA
5-7 January 2024

New York, New York, USA
26-30 June 2024

ISBN: 979-8-3313-1937-3

Printed from e-media with permission by:

Curran Associates, Inc.
57 Morehouse Lane
Red Hook, NY 12571

Labor and Employment Relations Association (LERA)
University of Illinois at Urbana-Champaign
121 Labor and Employment Relations Bldg.,
504 E. Armory Ave.
Champaign IL 61820
USA

Phone: (217) 333-0072
Fax: (217) 265-5130

LERAoffice@illinois.edu

Additional copies of this publication are available from:

Curran Associates, Inc.
57 Morehouse Lane
Red Hook, NY 12571 USA
Phone: 845-758-0400
Fax: 845-758-2633
Email: curran@proceedings.com
Web: www.proceedings.com

LABOR AND EMPLOYMENT RELATIONS ASSOCIATION SERIES

Proceedings of the LERA 2024 Meetings

LERA@ASSA Meeting
January 5–7, 2024
San Antonio, TX

LERA 76th Annual Meeting (2024 Triad)
June 26–30, 2024
New York, NY

OFFICERS OF LERA FOR 2023–2024

PRESIDENT
Dennis Dabney, Dabney Law LLC

PRESIDENT ELECT
Jim Pruitt, Kaiser Permanente

PAST PRESIDENT
Paul F. Clark, Pennsylvania State University

SECRETARY-TREASURER
Andrew Weaver, University of Illinois Urbana-Champaign

EDITOR-IN-CHIEF
J. Ryan Lamare, University of Illinois Urbana-Champaign

CHAPTER ADVISORY COUNCIL CHAIR
William Canak, Middle Tennessee State University (ret.)

EXECUTIVE DIRECTOR
Emily Smith, University of Illinois Urbana-Champaign

REGIONAL VICE PRESIDENTS
Jim Pruitt, Kaiser Permanente (West Region)
Beverly E. Harrison, Arbitrator/Mediator (East Region)
Robert Chiaravalli, Strategic Labor and HR LLC (Mid Region)

EXECUTIVE BOARD MEMBERS
Peter Berg, Michigan State University
Sharon Block, Harvard Law
Andrea Cáceres, SHARE/AFSCME
Robert Chiaravalli, Strategic Labor and HR LLC
Julie Farb, AFL-CIO
Eliza Forsythe, University of Illinois Urbana-Champaign
Ruben Garcia, University of Nevada, Las Vegas
Janet Gillman, Oregon Employment Relations Board
Beverly Harrison, Arbitrator/Mediator
John Johnson, Southeastern Pennsylvania Transportation Authority
Glenard Middleton, AFSCME MD Council 67*
Deborah Moore-Carter, City of Baltimore
Deborah Mueller, CSEA Local 1000 AFSCME
Dionne Pohler, University of Saskatchewan
Jim Pruitt, Kaiser Permanente
Christine Riordan, University of Illinois Urbana-Champaign
Sean Rogers, University of Rhode Island
Christy Yoshitomi, Federal Mediation and Conciliation Service

LEGAL COUNSEL
Steven B. Rynecki, von Briesen, Purtell & Roper, S.C.

PAST PRESIDENTS OF LERA

1948—Edwin E. Witte
1949—Sumner H. Slichter
1950—George W. Taylor
1951—William Leiserson
1952—J. Douglas Brown
1953—Ewan Clague
1954—Clark Kerr
1955—Lloyd G. Reynolds
1956—Richard A. Lester
1957—Dale Yoder
1958—E. Wight Bakke
1959—William Haber
1960—John T. Dunlop
1961—Philip Taft
1962—Charles A. Myers
1963—William F. Whyte
1964—Solomon Barkin
1965—Edwin Young
1966—Arthur M. Ross
1967—Neil Chamberlain
1968—George P. Shultz
1969—Frederick Harbison
1970—Douglass V. Brown
1971—George Hildebrand
1972—Benjamin Aaron
1973—Douglas H. Soutar
1974—Nathaniel Goldfinger
1975—Gerald G. Somers
1976—Irving Bernstein
1977—F. Ray Marshall
1978—Charles C. Killingsworth
1979—Jerome M. Rosow
1980—Jack Barbash
1981—Rudolph A. Oswald
1982—Milton Derber
1983—Jack Stieber
1984—Wayne L. Horvitz
1985—Everett Kassalow
1986—Lloyd Ulman
1987—Michael H. Moskow
1988—Phyllis A. Wallace
1989—Joyce D. Miller
1990—Robert B. McKersie
1991—James L. Stern
1992—Ernest J. Savoie
1993—George Strauss
1994—Lynn R. Williams
1995—Walter Gershenfeld
1996—Hoyt N. Wheeler
1997—Francine D. Blau
1998—F. Donal O'Brien
1999—Thomas A. Kochan
2000—Sheldon Friedman
2001—Magdalena Jacobsen
2002—John F. Burton, Jr.
2003—Paula B. Voos
2004—Marlene Heyser
2005—Stephen R. Sleigh
2006—David B. Lipsky
2007—Eileen B. Hoffman
2008—Anthony Oliver, Jr.
2009—Joel Cutcher-Gershenfeld
2010—Eileen Appelbaum
2011—Gordon Pavy
2012/13—David Lewin
2013/14—Lisa Lynch
2014/15—Martin Mulloy
2015/16—Bonnie Castrey
2016/17—Janice Bellace
2017/18—Harry Katz
2018/19—Kris Rondeau
2019/20—Dennis Dabney
2020/21—Adrienne Eaton
2021/22—Wilma Liebman
2022/23—Paul F. Clark
2023—William E. Spriggs*

**Deceased in office*

LABOR AND EMPLOYMENT RELATIONS ASSOCIATION SERIES

Proceedings of the LERA 2024 Meetings

LERA@ASSA Meeting
January 5–7, 2024, San Antonio, TX
(in conjunction with ASSA/AEA)

and

LERA 76th Annual Meeting (2024 Triad)
June 26–30, 2024, New York, NY

J. Ryan Lamare, Editor-in-Chief

PROCEEDINGS OF THE LERA 2024 MEETINGS

First Edition

LABOR AND EMPLOYMENT RELATIONS ASSOCIATION SERIES
(Formerly the Industrial Relations Research Association)

Proceedings of the Annual Meeting
Annual Research Volume
Perspectives on Work Magazine

Inquiries and other communications regarding membership, meetings, publications, and general affairs of the Association, as well as notice of address changes, should be addressed to the LERA office.

LABOR AND EMPLOYMENT RELATIONS ASSOCIATION
University of Illinois at Urbana-Champaign School of Labor and Employment Relations
121 Labor and Employment Relations Building, MC-504
504 East Armory Avenue, Champaign, IL 61820
Telephone: 217/333-0072 Fax: 217/265-5130 E-mail: mailto:LERAoffice@illinois.edu
Website: https://lera.memberclicks.net

CONTENTS

VII. UNION POWER, SATISFACTION, AND COMPOSITION

Angela B. Cornell, Chair

VIII. EMERGING TRENDS IN THE WORKPLACE

Edward Martin Davidson, Presiding

IX. CARING FOR THE CAREGIVERS

Kathy Hall, Presiding

X. WORKING IN THE PLATFORM ECONOMY

Amir Jafar, Presiding

XI. INTERSECTIONALITY AND WORKER VOICE

Weihao Li, Presidings

XII. HEALTH AND SAFETY

Lucy Taksa, Presiding

XIII. CLOSE TO THE EDGE: THE WORKING POOR AND THE PRECARIOUSNESS OF EMPLOYMENT

XIV. 2024 LERA LIFETIME ACHIEVEMENT AWARDS

XV. LERA ANNUAL REPORTS

I. Presidential Address

A Better Way to Work

DENNIS DABNEY
Kaiser Permanente, retired

Good afternoon and welcome to the LERA 76th annual conference in conjunction with the FMCS and ILERA.

I would like to begin this address with an acknowledgment of our colleague William E. Spriggs, whose presidential term I have the honor of completing. I would ask that we take a brief moment to think of Bill and his family. Thank you.

I remember many Sundays our Baptist minister would say, "Don't worry, I am not going to be long." Thirty minutes later, he is still talking. Well, that won't be me. I have a very short message that is similar to my 2021 presidential address.

In that address, I proposed that positive union–management relationships are the way to go. But I think I neglected to say why that is my belief. Needless to say, I assumed everyone would conclude that it was related to my vast and varied human resource experience in the labor relations field.

Then I remembered a presentation that I would travel the country to make to anyone who had an interest in a better way to work together.

So today, I will give you the short version of that presentation and hope it provides some rationale for my belief that there is a better way to work for all parties involved in the employment relationship.

But this is the United States, so nothing comes without a commercial break. I want to introduce you to the Kaiser Permanente that I retired from in 2021. At that time, it was the leading nonprofit health plan and delivery system serving more than 12 million members with locations across the country. So, we had different geographical locations, personalities, and beliefs to work with.

The cornerstone of our relationship with labor was the labor–management partnership. It is the longest-lived and largest partnership in the country. The important key here is that it improves service to our patients and our quality of care, and it eliminates waste in the system. The partnership is a jointly led, shared strategy for change. It's a process for employee engagement, workplace innovation, and performance. How is that possible? Well, I am glad you asked.

It's our self-directed teams, called unit-based teams, that are the catalysts for many of the positive results for the employees and the employer. These teams are rated on a five-point scale of performance, and 80 percent of the KP teams at that time were at the highest level, which is five. That's the full collaboration and measurable success level.

Initially, I thought these are employer gains, and they definitely are valued by the employer. But then I thought when employees love their work environment, feel free to speak up, are safer, and have appreciative patients — that's good for them as well.

But the employer gets what I call a "two-fer" in this case. Bernard Tyson, the former CEO of Kaiser, described it better than anyone:

"The company now has a competitive advantage. That's a leg up on the competition. When employees are engaged, you get this visible boost that money can't buy."

I have seen it at KP and last year at our Detroit Motor City conference, as I engaged with various employees at the hotel.

Companies gain what I just described, as well as the added benefits of operational stability and membership growth. Engaged employees working on improvement across the organization. So, not only do employees and employers gain — but unions do as well.

Unions in the Kaiser case have also gained leading wages and benefits for their members, as well as income and employment security. The institutional union has gained organizing neutrality and growth.

Many others have recognized that there is a better way. So, at the end of the day, everyone can benefit instead of having winners and losers that flip back and forth based on the economy or politics. As the title of my presentation says, it's a better way to work.

If you are looking for more information about a better way, I refer you to a Brandeis class on negotiating labor agreements taught by six of your former presidents and others. You can find out about it at https://tinyurl.com/2ddbm2my.

In closing, I would like to recommend a book that describes a better way to manage your workplace: *Everybody Matters: The Extraordinary Power of Caring for Your People Like Family* by Bob Chapman and Raj Sisodia.

II. Can I Come In? Accommodations and Inclusion

Missing the Past? Exploring Pathways of Boomerang Employees

MOONYOUNG JANG
Korea University

Abstract

Boomerang employees—those who return to a previous employer after a period away—are an increasingly noteworthy phenomenon that presents both opportunities and challenges for organizations and individual employees. Companies benefit from rehiring boomerang employees due to their familiarity with the organizational culture and processes, as well as the fresh perspectives they bring, which can facilitate quicker reintegration and enhanced performance. For individuals, informal job searches through personal networks often provide updated information about their former employers, aiding their decision to return. This study addresses a critical gap by integrating affective events theory to explore the pathways of boomerang employees. Work shock emerged as the only significant independent variable, forming part of a full mediation model. This model links negative affect and job dissatisfaction to boomerang intention, which ultimately demonstrates a significant pathway. Another contribution of this study lies in its empirical investigation of variables previously examined through qualitative research. Furthermore, the study explores the underlying mechanisms of boomerang intention by combining theoretical frameworks from affective events theory and network theory.

Introduction

Some employees choose to return to their previous company for a variety of reasons. Rehiring former employees—referred to as boomerang employees—offers several advantages for companies. For the employees, informal job searches often provide updated information about their former employer, influencing their decision to return.

The phenomenon of boomerang employees has grown substantially on a global scale. LinkedIn data show that the proportion of boomerang employees among new hires increased from 3.9% in 2019 to 4.5% in 2021, reflecting broader labor market shifts influenced by the COVID-19 pandemic and the Great Resignation (ICAEW 2023; SHRM 2022;). In the United States, voluntary quit rates reached record highs in 2021, resulting in a larger pool of potential boomerang employees. A survey found that 76% of employees who resigned during the pandemic would consider returning to their former employers (Adecco Group 2023). Similarly, the United Kingdom has seen a significant rise in boomerang employees, with rates increasing from 1.9% in 2019 to 2.23% in 2022, particularly within the financial services sector. This trend is driven by employees' familiarity with organizational culture and their ability to swiftly reintegrate and contribute effectively (Adecco Group 2023; ICAEW 2023).

The concept of boomerang employees has attracted growing interest in both academic and professional circles. This phenomenon can be explained by several factors, including the benefits companies derive from rehiring former employees who are already familiar with the organizational culture and processes. These returning employees often bring updated knowledge and skills gained during their time away, which can lead to improved performance and quicker acclimatization compared to new hires (Shipp et al. 2014; Swider et al. 2017).

Previous research has examined various aspects of boomerang employees, including the reasons for their return and the benefits to organizations. However, a significant gap exists in the literature regarding the internal psychological mechanisms of these individuals, particularly the role of negative affect and job dissatisfaction resulting from career, work, or personal shocks. This study seeks to address this gap by integrating affective events theory (AET) to explore how these factors influence boomerang intentions (Weiss and Cropanzano 1996). Furthermore, it investigates the moderating effect of informal job search behaviors, which involve leveraging personal networks to obtain information about potential reemployment opportunities (Trimble and Kmec 2011).

By developing a comprehensive model that accounts for these pathways, this research aims to significantly enhance the understanding of boomerang employees, providing valuable insights for organizations seeking to reduce turnover and effectively manage re-recruitment strategies.

Several previous studies have examined boomerang employees (Keller et al. 2021; Maier et al. 2021; Tian et al. 2022; Van Hooft et al. 2021), but fewer have focused on the individual psychological factors, such as shocks and affects, or the moderating role of informal job searches in shaping their pathways. While some research has touched on these aspects, it remains fragmented, highlighting the need for an integrated model to comprehensively explore the pathways of boomerang employees. Although external conditions may not change significantly, employees often require updated information about their former company to decide whether to return. Informal job searches act as a moderating variable in boomerang intention, influencing the decision-making process for returning to a previous employer.

Existing research on informal job searches provides insights into the processes (e.g., the use of network contacts) and outcomes (e.g., the quality of subsequent employment). In situations involving job dissatisfaction, mediating factors may include an individual employee's negative affect, generated by various shocks (career, work, or personal). These factors link job dissatisfaction to a boomerang employee's intention, with informal job searches acting as a moderating factor. Affective events theory offers a strong foundation for understanding these relationships.

I expect this study to contribute to the literature on boomerang employees' intentions, which is important for protecting a company from employee turnover and encouraging the recruitment of high performers.

Informal job search refers to the intentional use of personal contacts, such as friends, family, or colleagues, to obtain information about or influence the acquisition of alternative employment (Trimble and Kmec 2011).

Affective events theory (Weiss and Cropanzano 1996) explains many things about employees' emotions leading to boomerang intention, with the mediation effect of negative affect and job dissatisfaction in the relationship between total shocks (career, work, and personal) and boomerang employees' intention.

The dual mediation model in this study includes the following processes: (1) whether there is shock (career, work, personal), (2) negative affect is generated with consideration of affective events theory, (3) job dissatisfaction occurs, and (4) the dependent variable is a boomerang employee's intention.

Figure 1. Hypothesized research model

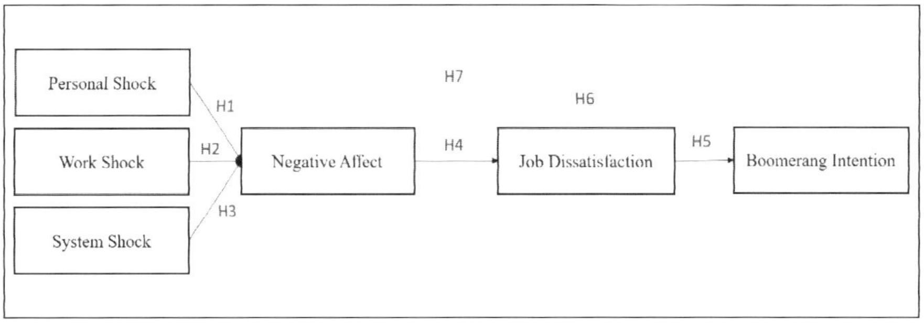

Literature Review

Voluntary Turnover

Turnover refers to the dynamic interaction between an individual and an organization, involving the stages of attachment, exchange, and separation (Mossholder, Settoon, and Henagan, 2005). The existing literature on turnover has primarily focused on two objectives: identifying the factors that lead personnel to leave an organization (i.e., the predictors of turnover) and understanding the manner in which they depart (i.e., the process of turnover) (Shipp et al. 2014). The researchers made two distinct references to the content and process models of turnover, as outlined by Maertz and Campion (2004). Voluntary turnover is particularly significant because it can impact an organization's performance in achieving its goals (Hom and Griffeth 1995).

Boomerang Intention

Keller et al. (2021) contend that boomerang employees possess familiarity with the organization's social system, which newly hired individuals lack. Social systems often undergo gradual evolution, with networks of shared knowledge and relationships playing a crucial role in establishing stability and acting as a potent force of resistance to change (Zander 1992). Hence, the boomerang employee's tacit knowledge, acquired during their earlier tenure at the organization, is expected to persist and serve as a valuable asset upon reemployment. Boomerang employees also possess superior skills in utilizing their ideas and contributions compared to new hires. This enables them to outperform new hires initially, as they are better able to coordinate with others and encounter less resistance when developing their ideas and contributions.

Shocks—Career Shock, Work Shock, Personal Shock

Shock, within the context of the job life cycle, refers to a sudden and disruptive incident that significantly contradicts an employee's values to the extent that they contemplate altering their conduct (Lee et al. 1999). Shocks, as seen by Morgeson et al. (2015) and Morgeson and DeRue (2006), have the ability to interrupt established patterns and impair automatic cognitive processing in employees. Encountering a shock disrupts the existing state of affairs. An unexpected event can lead employees to reassess their views on their current job (Holtom et al. 2012) and potentially intensify their inclination to resign (Wilson and Schooler 1991). There are two types of general shocks that have been identified in literature (Holtom et al. 2005): personal shocks and job shocks. Personal shocks arise from individual matters, particularly those pertaining to family, whereas professional shocks stem from the overall substance, character, and circumstances of one's employment inside an organization, such as a conflict with a supervisor or colleague, or the receipt of a job offer.

The Mediators

Negative Affects

Three items that asked respondents to indicate how often they felt excited or enthusiastic during the job search during the preceding three months (using 5-point scales; $\alpha = .73$). These items are indicators of the high activation dimension of positive affect and were adapted from the job affect scale (Burke et al. 1989). In this research, positive affects are excluded from survey items due to relevance. Examples of negative emotions are scared and distressed.

Job Dissatisfaction

Job dissatisfaction is commonly described as a negative emotional or attitudinal response to one's occupation (Spector 1997). Many turnover models incorporate job satisfaction as a key factor influencing turnover. These models distinguish between job dissatisfaction and sudden changes as separate aspects of turnover. More specifically, there is a direct correlation between increased work unhappiness and elevated turnover rates (Hom and Kinicki 2001; Lee 1988; Zimmerman and Darnold 2009). This research postulates that work unhappiness can be a motivating factor for employees to consider leaving their current positions.

My hypotheses are as follows:

- Hypothesis 1 (H1). System shock positively influences negative affect and job dissatisfaction on intention to boomerang.

- Hypothesis 2 (H2). Work shock positively influences negative affect and job dissatisfaction on intention to boomerang.
- Hypothesis 3 (H3). Personal shock positively influences negative affect and job dissatisfaction on intention to boomerang.
- Hypothesis 4 (H4). Negative affect positively influences job dissatisfaction on intention to boomerang.
- Hypothesis 5 (H5). Job dissatisfaction positively influences intention to boomerang.
- Hypothesis 6 (H6). Job dissatisfaction mediates the relationship between negative affect and intention to boomerang.
- Hypothesis 7 (H7). Negative affect and job dissatisfaction dually mediate the relationship among system, work, and personal shock and on intention to boomerang.

Theoretical Backgrounds

Affective Events Theory

According to affective events theory (AET), developed by Weiss and Cropanzano (1996), emotional experiences in the workplace play a crucial role in influencing job satisfaction and subsequent behaviors. AET diverges from traditional models of job satisfaction by focusing on the organization, origins, and outcomes of these emotional experiences rather than solely on cognitive evaluations.

AET posits that workplace events are immediate triggers for emotional responses, which then affect attitudes and behaviors. Traditional models emphasize stable environmental characteristics, while AET suggests that these characteristics indirectly influence emotions by increasing or decreasing the likelihood of affective events. This perspective highlights the importance of time, noting that emotional experiences and their impacts fluctuate, something that traditional models often overlook by considering job satisfaction at a single point in time (Weiss and Cropanzano 1996).

Key components of AET include the distinction between moods and emotions, with emotions being more transient and event specific. The theory identifies both endogenous factors (e.g., mood cycles) and exogenous factors (e.g., significant workplace events) as contributors to affective variations. Emotional experiences can lead to two types of behaviors: affect-driven behaviors, which are immediate and not mediated by overall job attitudes, and judgment-driven behaviors, which are influenced by cognitive evaluations of job satisfaction.

Overall, AET provides a comprehensive framework for understanding the dynamic interplay among emotional events, job satisfaction, and workplace behaviors, emphasizing the multifaceted nature of affective experiences. The theory offers a nuanced perspective that complements and extends traditional models by integrating the temporal patterns and organizational context of emotional experiences in the workplace.

Methods and Results

Methods

I adopted structural equation modeling (SEM) to analyze the relationships between variables derived from self-report surveys. Data were collected from a sample of 300 participants using an online survey administered by Embrain, a reputable survey company known for its comprehensive data collection services. The self-report surveys included a variety of measures designed to capture the constructs of interest. SEM was chosen for its robustness in testing complex relationships and its ability to handle latent variables, providing a comprehensive understanding of the underlying theoretical model. The data collection and analysis procedures adhered to rigorous standards to ensure the validity and reliability of the findings.

The collected data were analyzed using programs such as SPSS 26.0 and SmartPLS 4, with career shock, work shock, and personal shock as independent variables. After exploratory factor analysis, the variables were reclassified as personal, work, and system shocks. Partial least squares (PLS)-SEM was used as the methodology to analyze the model, with negative affect and job dissatisfaction as mediating factors and intention to boomerang as the dependent variable. Exploratory and confirmatory factor analysis, discriminant validity analysis, and reliability analysis were conducted. Direct effects, detailed indirect effects, and total effects were calculated, and 90% confidence intervals (CIs) were computed by bootstrapping 5,000 times to determine significance. While 95% CI is typically used for bootstrapping, the analysis revealed that paths significant at the 0.05 level were not significant. To address this issue, the range was widened to 90% CI to ensure significance.

Results

Demographic Distribution

The demographic distribution of the study participants is shown in Table 1. The survey was conducted on employees who changed jobs, and 300 samples were obtained. Of the total 270 respondents who answered the questions about full-time employment status and total years of service, 39.3% were male and 60.7% were female. In terms of age, 14.8% were between 19 and 29 years old, 45.9% were between 30 and 39 years old, 30.0% were between 40 and 49 years old, and 9.3% were between 50 and 59 years old. Regarding job title, 4.1% were temporary employees, 60.7% were employees/agents, 33.7% were managers/assistant managers/ directors, and 1.5% were executives. As for company size, 63.3% worked for small businesses, 25.6% for medium-sized businesses, and 11.1% for large businesses. In terms of industry, 0.7% worked in agriculture/fishing/forestry, 14.1% in light industry/heavy industry, 48.1% in services, and 37.0% in other industries. Additionally, 19.1% had been with their last company for less than one year, 45.2% for one to five years, 24.4% for five to ten years, and 12.2% for more than ten years. Regarding current company tenure, 3.7% had been with their company for less than one year, 31.9% for one to less than five years, 31.1% for five to less than ten years, and 33.3% for more than ten years.

Table 1. Demographic distribution

Variable	Item	Frequency	Percentile
Gender	Male	106	39.3
	Female	164	60.7
Age range	19 to 29 years old	40	14.8
	30 to 39 years old	124	45.9
	40 to 49 years old	81	30.0
	50 to 59 years old	25	9.3
Roles	Temporary jobs	11	4.1
	Employee/associate	164	60.7
	Deputy/director	91	33.7
	Executive level	4	1.5
Scale	Small and medium-sized businesses	171	63.3
	Midsize businesses	69	25.6
	Large enterprises	30	11.1

Industries	Agriculture/fishing/forestry	2	0.7
	Light/heavy chemical	38	14.1
	Hospitality	130	48.1
	Other	100	37.0
Length of service at current employer	Less than 1 year	49	18.1
	More than 1 year to less than 5 years	122	45.2
	More than 5 years to less than 10 years	66	24.4
	10 years or more	33	12.2
Overall tenure	Less than 1 year	10	3.7
	More than 1 year to less than 5 years	86	31.9
	More than 5 years to less than 10 years	84	31.1
	10 years or more	90	33.3
Total		**270**	**100.0**

Preliminary Analysis

HTMT (Heterotrait-Monotrait Ratio) Analysis

Table 2. Discriminant validity: Heterotrait-monotrait ratio (HTMT), matrix

	Preliminary Analysis	Preliminary Analysis	Preliminary Analysis	Preliminary Analysis	Preliminary Analysis	Preliminary Analysis
BI						
JDS	0.252					
NegA	0.111	0.656				
Personal	0.278	0.103	0.185			
System	0.201	0.274	0.188	0.590		
Work	0.300	0.195	0.192	0.617	0.864	

With another method for testing the discriminant validity—an HTMT-ratio approach—HTMT-ratio values less than 0.90 are considered acceptable. Table 2 shows that all values were below 0.9, which indicates that the discriminant validity was well maintained between constructs.

VIF (Variance Inflation Factor)

Table 3. Variance inflation factor (VIF), outer model, list

	VIF
bi1	4.232
bi2	5.213
bi3	4.279
c_shock1	1.410
c_shock2	1.215
c_shock3	2.142
c_shock4	1.807
dissat1	1.508
dissat2	1.458
dissat3	1.402
dissat5	1.273
nega1	3.826
nega2	3.443
nega3	3.104
nega4	4.424
nega5	3.544
p_shock2	1.703
p_shock3	2.060
p_shock4	1.625
w_shock1	1.182
w_shock2	1.284
w_shock3	1.416
w_shock4	1.361

Before evaluation of the structural paths of the research model, a diagnosis of multicollinearity was conducted on a set of predictive constructs. In general, a threshold of below 5 for VIF is recommended (Hair et al. 2022). As can be seen in Table 4, all values are lower than 5, so multicollinearity between predictive constructs was not a problem in the structural model.

Table 4. Variance inflation factor (VIF), inner model, matrix

	BI	JDS	NegA	Personal	System	Work
BI						
JDS	1.482					
NegA	1.496	1.051				
Personal	1.347	1.326	1.321			
System	1.695	1.650	1.650			
Work	1.748	1.746	1.717			

Exploratory Factor Analysis

In the exploratory factor analysis (EFA) conducted using SPSS version 26.0, the Kaiser–Meyer–Olkin test and Bartlett's test of sphericity showed a significance probability of 0.000. The total variance explained by the three factors was 64.021%, with cumulative squared loadings of 64.021%, indicating strong explanatory power (over 60%). Based on the three factors extracted through principal component analysis and Varimax rotation, I reorganized the factors—personal, work, and career shocks—into three categories: personal, work, and system.

Table 5. Exploratory factor analysis, rotated factors matrix

	Factors		
	1	**2**	**3**
P_Shock3	.837	.195	.106
P_Shock2	.835	.065	.201
P_Shock4	.816	.238	.073
C_Shock3	.119	.877	.154
C_Shock4	.147	.848	.022
W_Shock3	.285	.582	.186
C_Shock1	.123	.580	.389
W_Shock4	.103	.497	.411
W_Shock2	.063	.115	.830
W_Shock1	.079	.150	.744
C_Shock2	.253	.410	.513

Principal component analysis (PCA).

Rotation method: Varimax rotation with Kaiser normalization (rotation converged in 5 iterations).

CFA (Confirmatory Factor Analysis)

I examined how well the observed variables that make up the latent variable explain it through factor rotation. I found that all loadings exceeded 0.5. The results of a 5,000-sample bootstrapping method show that the latent variable indicators for each observed variable are significant, as the p-value is less than 0.05 or the 95% CI does not include zero in the bias-corrected (BC) bootstrapping analysis.

The criteria for indicators applied in confirmatory factor analysis are as follows. First, regarding reliability, a Cronbach's α of 0.7 or higher is considered good for internal consistency, and the reliability of the indicator, in terms of factor loadings, is ideally 0.7 or higher. This is because, for the average variance extracted (AVE) value of a latent variable to be 0.5 or higher, the factor loadings of the observed variables that comprise the latent variable must average 0.7 or higher. However, if the average is 0.7 or higher, it is acceptable for some items to have factor loadings of 0.5 or higher. For construct validity, the significance of the observation should be 0.05 or less based on the p-value, or 1.96 or higher based on the t-value. Bootstrapping can also be used to determine whether the 95% confidence interval includes zero. An AVE index of 0.5 or higher is a common standard, though it can be slightly lower as long as discriminant validity is maintained. Discriminant validity requires that the square root of the AVE be higher than the correlation coefficient between the latent variable and other latent variables. Otherwise, redundant latent variables should be removed or other highly correlated observed variables combined into a single factor for analysis.

Table 6. Confirmatory factor analysis

Paths	Original Outer Loadings	Bootstrapping			
		Mean	SD	t	p
bi1 – BI	0.944	0.944	0.009	110.593	0.000
bi2 – BI	0.956	0.956	0.010	98.534	0.000
bi3 – BI	0.943	0.942	0.014	65.210	0.000
p_shock2 - Personal	0.850	0.847	0.038	22.153	0.000
p_shock3 - Personal	0.901	0.899	0.032	27.775	0.000
p_shock4 - Personal	0.778	0.773	0.053	14.741	0.000
w_shock1 - System	0.740	0.726	0.105	7.045	0.000
w_shock2 - System	0.726	0.715	0.102	7.086	0.000
c_shock2 - System	0.779	0.767	0.083	9.402	0.000
c_shock1 - Work	0.770	0.754	0.081	9.461	0.000
c_shock3 - Work	0.701	0.683	0.124	5.673	0.000
c_shock4 - Work	0.608	0.592	0.139	4.369	0.000
w_shock3 - Work	0.683	0.664	0.103	6.634	0.000
w_shock4 - Work	0.762	0.746	0.083	9.134	0.000
dissat1 – JDS	0.771	0.769	0.037	20.729	0.000
dissat2 – JDS	0.786	0.784	0.032	24.883	0.000
dissat3 – JDS	0.744	0.741	0.040	18.484	0.000
dissat5 – JDS	0.689	0.686	0.052	13.342	0.000
nega1 – NegA	0.859	0.858	0.019	45.848	0.000
nega2 – NegA	0.902	0.901	0.012	73.477	0.000
nega3 – NegA	0.819	0.819	0.021	38.680	0.000
nega4 – NegA	0.896	0.895	0.013	66.946	0.000
nega5 – NegA	0.847	0.846	0.018	46.884	0.000

Table 7. Criteria for external model/reflective indicators

Separation	Item	Criteria	References
Reliability	Internal consistency	Cronbach's α ≥ 0.7	Nuunally and Bernstein (1994)
		Composite reliability ≥ 0.7	Bogozzi and Yi (2016)
	Metric reliability	Factor loadings of the measurement factors ≥ 0.7 (0.6 degrees is acceptable for exploratory studies)	Chin (1998b) Hair et al. (2012b)
Convergent validity	Significance of metrics	t-value 1.96 or p 0.05 (estimated using bootstrap)	Gofen-Straub (2005)
	Average variance extracted (AVE)	AVE ≥ 0.5	Fornell and Larcker (1981)
Discriminant validity	Cross-loading criteria	In confirmatory factor analysis, measures should load distinctly higher on the latent variable they measure than on other latent variables	Gofen-Straub (2005)
	Gofen-Straub Criteria	The root mean square value of the average variance extracted (AVE) is higher than the correlation coefficient between the latent variable and the other latent variables	Gofen-Straub (2005)

Based on these criteria, here are the results of the study's analysis: First, the internal consistency (Cronbach's alpha) of the observed variables that make up the latent variables is above 0.7 for all except system shock. Additionally, the composite reliability was found to be greater than 0.7, and the AVE exceeded 0.5. Therefore, it was confirmed that the latent variables can be used to create a research model and test hypotheses.

Regarding model fit, unlike CB-SEM, partial least squares structural equation modeling (PLS-SEM) analysis has not yet established various goodness-of-fit criteria. The goodness of fit provided by the program is shown in Table 5. The SRMR of the estimated model is 0.075, which satisfies the general goodness-of-fit criterion of less than 0.08 (Hu and Bentler 1999). Therefore, the model can be considered suitable for analysis.

To analyze the discriminant validity of the latent variables in the structural equation model, the Fornell–Larcker criterion is used. The diagonal values represent the square root of the AVE, while the off-diagonal values represent the correlation coefficients between the two latent variables. Discriminant validity is confirmed if the correlation coefficient between two latent variables is less than the square root of the AVE for each latent variable. Based on these results, it is confirmed that the research model can be analyzed using the latent variables derived from the observed variables.

Table 8. Construct reliability and validity

	Cronbach's Alpha	**Composite Reliability (rho_a)**	**Average Variance Extracted (AVE)**
BI	.943	0.943	.898
JDS	.738	0.743	.560
NegA	.916	0.926	.748
Personal	.801	0.833	.714
System	.613	0.618	.560
Work	.771	0.789	.500

Table 9. Model fit

	Estimated Model	**Criteria**
SRMR	.075	< 0.1 or < 0.08 in conservative version
Chi-square	867.358	

Hu and Bentler (1999).

Table 10. Latent variable discriminant validity by Fornell–Larcker

	BI	JDS	NegA	Personal	System	Work
BI	.947					
JDS	.207	.748				
NegA	.110	.549	.865			
Personal	.251	.024	.154	.845		
System	.159	.185	.149	.422	.749	
Work	.255	.132	.210	.458	.606	.707

Diagonal cells are the square root of AV; nondiagonal cells are the correlation coefficient between the two latent variables.

Analysis of SEM (Structural Equation Model)

The direct effects of the structural model analysis are presented in Table 7 and Figure 2. Table 6 displays the original coefficient estimates along with the mean and standard deviation of the estimates derived from 5,000 bootstrapping runs, indicating that the two sets of coefficients are consistent. In general, bootstrapping is used to test the significance of coefficients. A coefficient is considered significant if the p-value is 0.05 or less, or if the 90% CI does not include zero. The results show that all direct effect paths are significant, with the 90% CIs excluding zero, thereby confirming their significance.

The direct effects of the paths between each latent variable, as calculated by standardized coefficients, are as follows. First, we find that personal shocks have a significant defining effect on intentions to boomerang, with β = .201 (90% CI: .089~.305). Work shock has a significant defining effect of β = .194 (90% CI: .025~.316) on boomerang intention, and institutional shock has a significant defining effect of β = .173 (90% CI: .057~.276) on job dissatisfaction. Furthermore, job dissatisfaction has a significant defining effect on boomerang intention with β = .233 (90% CI: .101~.347), and negative emotions have a significant defining effect on boomerang intention with β = .548 (90% CI: .460~.624).

These results suggest that the stronger the personal or work shock, the higher the boomerang intention, and the stronger the institutional shock, the higher the job dissatisfaction. In turn, higher job dissatisfaction is associated with higher boomerang intentions, and stronger negative emotions are associated with higher job dissatisfaction.

Table 11. Direct effects

	Original Coefficient	Bias-Corrected (BC) Bootstrapping					
						90% CI	
		Mean	SD	t	p	Lower	Upper
JDS – BI	**0.233**	**0.233**	**0.074**	**0.074**	**0.002**	**0.101**	**0.347**
NegA – BI	-0.079	-0.079	0.075	0.075	0.292	-0.197	0.053
NegA – JDS	**0.548**	**0.548**	**0.050**	**0.050**	**0.000**	**0.460**	**0.624**
Personal – BI	**0.201**	**0.199**	**0.066**	**0.066**	**0.003**	**0.089**	**0.305**
Personal – JDS	**-0.118**	**-0.117**	**0.064**	**0.064**	**0.066**	**-0.222**	**-0.012**
Personal – NegA	0.070	0.074	0.076	0.076	0.356	-0.056	0.193
System – BI	-0.075	-0.064	0.081	0.081	0.353	-0.217	0.050
System – JDS	**0.173**	**0.175**	**0.067**	**0.067**	**0.010**	**0.057**	**0.276**

System – NegA	0.019	0.026	0.087	0.087	0.827	-0.129	0.156
Work – BI	**0.194**	**0.197**	**0.088**	**0.088**	**0.027**	**0.025**	**0.316**
Work – JDS	-0.034	-0.029	0.072	0.072	0.635	-0.155	0.076
Work – NegA	0.166	0.168	0.114	0.114	0.144	-0.064	0.316

JDS - BI: A positive coefficient (0.233) indicates a significant positive direct effect of JDS on BI ($p < 0.05$).

NegA - BI: A negative coefficient (–0.079) suggests a negative, but not statistically significant, effect of NegA on BI.

NegA - JDS: A strong positive and significant effect (0.548) of NegA on JDS ($p < 0.05$).

Personal - BI: A positive and significant effect (0.201) of personal shocks on BI ($p < 0.05$).

Personal - JDS: A negative coefficient (–0.118) suggests a negative direct effect of personal shocks on JDS ($p < 0.10$).

Personal - NegA: A small positive effect (0.070) of personal shocks on NegA, but not statistically significant.

System - BI: A negative coefficient (–0.075) suggests a nonsignificant negative effect of system shocks on BI.

System - JDS: A positive and significant effect (0.173) of system shocks on JDS ($p < 0.05$).

System - NegA: A nonsignificant positive effect (0.019) of system shocks on NegA.

Work - BI: A positive and significant effect (0.194) of work shocks on BI ($p < 0.05$).

Work - JDS: A negative coefficient (–0.034) suggests a nonsignificant negative effect of work shocks on JDS.

Work - NegA: A positive and potentially significant effect (0.166) of work shocks on NegA.

Figure 2. Factor loadings and structural path coefficients (standardized coefficients)

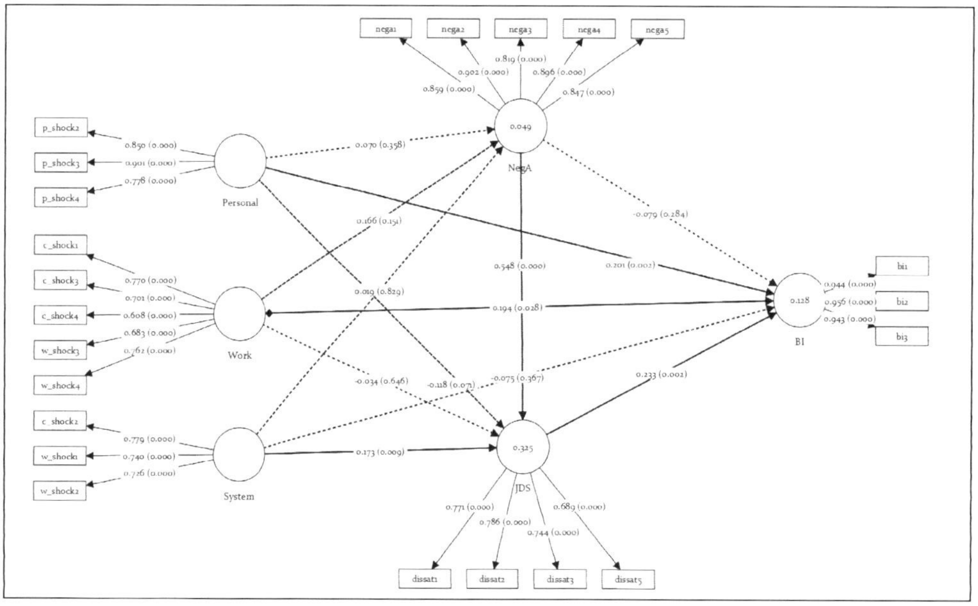

NegA (negative affects): Appears to be an important mediator variable that influences JDS (likely job dissatisfaction) and BI (boomerang intentions).

JDS (job dissatisfaction): Plays a crucial role in predicting BI.

Personal, work, system shocks: These shocks are shown to influence NegA, JDS, and BI through various paths, indicating that they are significant predictors in this model.

The indirect effects are shown in Table 12. First, we find that system shock has a significant definitional indirect effect on boomerang intention via job dissatisfaction with $\beta = .040$ (90% CI: .015~.077), and negative emotion has a significant definitional indirect effect on boomerang intention via job dissatisfaction with $\beta = .128$ (90% CI: .060~.201).

These results imply that in our model, negative emotions and job dissatisfaction do not significantly mediate the process through which shock-related variables affect boomerang intentions; instead, institutional shock and negative emotions significantly mediate boomerang intentions through job dissatisfaction.

Table 12. Indirect effects

Indirect Paths	Original Coefficient	BC Bootstrapping Mean	SD	t	p	90% CI Lower	90% CI Upper
NegA - JDS – BI	0.128	0.128	0.128	2.988	0.003	0.060	0.201
Personal - NegA - BI	–0.006	–0.005	–0.005	0.595	0.552	–0.030	0.003
Personal - JDS - BI	–0.028	–0.027	–0.027	1.516	0.129	–0.066	–0.004
Personal - NegA - JDS	0.039	0.040	0.040	0.921	0.357	–0.031	0.106
System - NegA - BI	–0.001	–0.002	–0.002	0.158	0.875	–0.023	0.009
System - JDS - BI	0.040	0.039	0.039	2.191	0.029	0.015	0.077
System - NegA - JDS	0.010	0.014	0.014	0.218	0.828	–0.072	0.085
Work - NegA - BI	–0.013	–0.014	–0.014	0.714	0.476	–0.055	0.006
Work - JDS – BI	–0.008	–0.007	–0.007	0.451	0.652	–0.043	0.016
Work - NegA - JDS	0.091	0.092	0.092	1.436	0.151	–0.030	0.179
Work - NegA - JDS - BI	0.021	0.021	0.021	1.268	0.205	–0.002	0.051
Personal - NegA - JDS - BI	0.009	0.010	0.010	0.836	0.403	–0.005	0.030
System - NegA - JDS - BI	0.002	0.003	0.003	0.208	0.835	–0.016	0.022

Total effects, which is the sum of the direct and indirect effects analyzed above, are shown in Table 13.

First, we find that job dissatisfaction has a significant definition effect on boomerang intention with a total effect of β–.233 (90%CI: .101~.347), negative emotions have a definition effect on job dissatisfaction with a total effect of β=.548 (90%CI: .460~.624), personal shocks have a definition effect with a total effect of β=.176 (90%CI: .064~.280) as a total effect, institutional shock has a definition effect on job dissatisfaction with β=.184 (90%CI: .039~.305) as a total effect, and finally, work shock has a significant definition effect on boomerang intention with β=.194 (90%CI: .031~.321) as a total effect.

Table 13. Total effects

Total Effect Path	Coefficient	BC Bootstrapping Mean	SD	t	p	90% CI Lower	90% CI Upper
JDS – BI	0.233	0.233	0.074	3.142	0.002	0.101	0.347
NegA – BI	0.049	0.049	0.067	0.737	0.461	–0.061	0.160
NegA – JDS	0.548	0.548	0.050	11.035	0.000	0.460	0.624
Personal - BI	0.176	0.176	0.066	2.676	0.007	0.064	0.280
Personal - JDS	–0.080	–0.076	0.079	1.009	0.313	–0.211	0.048

Personal - NegA	0.070	0.074	0.076	0.923	0.356	–0.056	0.193
System – BI	–0.033	–0.023	0.080	0.417	0.677	–0.178	0.085
System - JDS	0.184	0.188	0.081	2.258	0.024	0.039	0.305
System - NegA	0.019	0.026	0.087	0.218	0.827	–0.129	0.156
Work – BI	0.194	0.198	0.087	2.236	0.025	0.031	0.321

Discussion and Limitations

The structural equation model provides a comprehensive analysis of how various personal, work-related, and systemic shocks influence negative affect (NegA), job dissatisfaction (JDS), and boomerang intentions (BI). This analysis offers valuable insights into the dynamics of employee behavior, particularly regarding job dissatisfaction and the inclination to return to a former employer—a phenomenon known as boomerang intentions.

A key finding is the significant positive relationship between job dissatisfaction and boomerang intentions, with a coefficient of 0.233 that is statistically significant at the $p < 0.05$ level. This suggests that higher levels of job dissatisfaction may not only drive employees to leave but also increase the likelihood of them considering a return to their former employer. This counterintuitive finding underscores a complex psychological mechanism in which dissatisfaction, rather than entirely severing ties with an organization, may leave room for reconsideration and potential return—particularly under conditions such as improved work environments or changes in personal circumstances after departure.

Negative affect (NegA) plays a crucial role in this model, exhibiting a strong and significant positive effect on job dissatisfaction (coefficient = 0.548, $p < 0.05$). This robust relationship suggests that employees experiencing higher levels of negative emotions are more likely to feel dissatisfied with their jobs. This finding reflects how emotional states directly influence job perceptions, potentially intensifying feelings of discontent. However, the direct effect of NegA on boomerang intentions is negative (–0.079) and not statistically significant. This indicates that while negative emotions significantly contribute to job dissatisfaction, they do not directly drive the decision to return to a former employer. The nonsignificant relationship suggests that although negative emotions may sour current job experiences, the decision to return to a previous employer is more likely shaped by other factors, such as improved personal circumstances or changes within the former organization.

The role of personal shocks in this model is also significant. Personal shocks have a positive influence on boomerang intentions (coefficient = 0.201, $p < 0.05$), indicating that individual life events or changes outside of work can substantially impact an employee's decision to return to a former employer. This finding highlights the interplay between personal life and career decisions, suggesting that employees may reconsider past employment when faced with new personal circumstances that make returning more appealing or necessary. Conversely, personal shocks negatively affect job dissatisfaction (coefficient = –0.118, $p < 0.10$), implying that certain personal factors or life events may alleviate job dissatisfaction. This relationship could reflect a reevaluation of job dissatisfaction in light of external personal challenges, where the relative importance of workplace dissatisfaction decreases when significant personal issues arise.

System shocks, representing changes or disruptions in the organizational environment, have a significant positive effect on job dissatisfaction (coefficient = 0.173, $p < 0.05$). This finding suggests that employees are sensitive to changes in organizational structure or policies, which can lead to increased dissatisfaction. However, system shocks do not significantly impact either boomerang intentions or negative affect, indicating that while organizational changes can undermine job satisfaction, they do not necessarily evoke strong emotional responses or influence the decision to return to a previous employer. This may imply that employees perceive system shocks as an inevitable aspect of organizational life, affecting their satisfaction but not necessarily shaping their long-term career decisions.

Work shocks, defined as changes or disruptions related to the nature of the work itself, significantly influence boomerang intentions (coefficient = 0.194, $p < 0.05$). This finding suggests that changes in job roles, responsibilities, or work conditions are critical factors in employees' decisions to return to a former employer. However, work shocks do not significantly affect job dissatisfaction or negative affect, which may indicate that, while these changes can trigger career reconsiderations, they do not inherently make employees more dissatisfied or emotionally distressed. Instead, such shocks may prompt a pragmatic reevaluation of past employment options in light of current work challenges.

In conclusion, this model sheds light on the intricate relationships between personal, work, and system shocks and their effects on negative affect, job dissatisfaction, and boomerang intentions. Job dissatisfaction emerges as a key mediator, linking negative emotions and various shocks to the likelihood of returning to a previous employer. These findings underscore the importance of managing job satisfaction and understanding the nuanced effects of different types of shocks on employee behavior. By addressing the factors that contribute to dissatisfaction and understanding the circumstances under which employees consider returning to a former employer, organizations can better manage employee retention and reengagement strategies, ultimately fostering a more stable and committed workforce.

Additionally, incorporating variables related to boomerang intentions—such as legacy identification or psychological contract violation, which were highlighted in previous studies discussed in this article—could enhance the current research model. Doing so may yield additional contributions and provide deeper insights, especially as the model in its current form appears somewhat conventional.

Another contribution of this study lies in its empirical findings on variables previously examined through qualitative research. Additionally, this study contributes by exploring the underlying mechanism of boomerang intentions, offering theoretical frameworks that integrate affective events theory.

While numerous studies have examined boomerang employees, there is a lack of research focusing on the specific characteristics of individuals that lead to job dissatisfaction. This study addresses this gap by utilizing affective events theory and social network theory, exploring the mediating effects of negative affect and job dissatisfaction, and considering the moderating role of informal job searches. Although existing studies provide some insights, they remain incomplete. An integrated model is needed to fully explore the pathways of boomerang employees. However, limitations in available samples and insufficient information regarding processes and pathways persist.

The findings of this study offer a detailed understanding of the factors influencing boomerang intentions among employees. The results indicate that work-related shocks significantly affect negative affect and job dissatisfaction, which subsequently influence the intention to return to a former employer. This aligns with the broader literature emphasizing the role of context-specific factors in shaping employee behavior and turnover decisions (Holtom et al. 2008).

This study's limitations include reliance on self-reported data, which may be subject to biases, and its focus on a specific sample, potentially limiting the generalizability of the findings. Future research should aim to include more diverse samples and consider longitudinal designs to capture the dynamics of boomerang behavior over time. Additionally, exploring the impact of organizational culture and external economic conditions on boomerang intentions could provide further insights into the factors driving this phenomenon (Maertz and Campion 2004; Mitchell et al. 2001).

Theoretical Implications

According to the data, the variables for shocks (personal, work, career) were reclassified into personal, work, and system shocks following exploratory factor analysis. Work shocks primarily involved internal conflicts experienced by employees within the organization (e.g., "A mentor or colleague who is important to you has left the organization," "You successfully took on a new task or tangible project that initially challenged you in a new and significant way," "You received a pay raise, promotion, or desirable increase in responsibility sooner than expected," "My colleague was performing better than me," and "I had a clash with my supervisor due to organizational issues"). Additionally, system shocks were more closely associated with issues related to the organizational system (e.g., "No further education opportunities to deal with new trends," "My current employer does not pay me the salary I want to have," and "The organization experienced a significant negative event, such as a workforce reduction, bankruptcy, or major ethical scandal").

This study also extends the application of affective events theory (AET) to the context of boomerang employees. By demonstrating how negative affect and job dissatisfaction mediate the relationship between shocks (system, work, and personal) and boomerang intention, this research offers a more nuanced understanding of the emotional and cognitive processes that influence an employee's decision to return to a former employer, incorporating individual psychological variables (Granovetter 1973; Weiss and Cropanzano 1996).

In understanding boomerang intention, this study's findings contribute to the literature on employee turnover as well as to boomerang employees in particular by highlighting the specific conditions under which former employees are likely to return to their previous employers. This adds depth to existing models of voluntary turnover by incorporating the role of emotional responses to job-related shocks (Lee et al. 2004; Shipp et al. 2014).

The dual mediation model proposed in this study, in which negative affect and job dissatisfaction jointly mediate the relationship between shocks and boomerang intention, provides a comprehensive framework for understanding employee turnover and reentry behaviors. This model can be further tested and refined across various organizational settings to validate its applicability (Hom and Kinicki 2001; Judge et al. 2006).

Practical Implications

Organizations should adopt proactive measures to anticipate and mitigate the impact of personal, work, and system shocks on employees. These measures could include developing flexible work arrangements, providing support during organizational changes, and equipping employees to manage personal shocks by offering resources such as counseling or financial planning. Effectively managing these shocks can help organizations maintain stability and reduce the likelihood of employee turnover.

Companies should carefully consider the reasons for an employee's initial departure before deciding to rehire them. While boomerang employees can bring valuable experience, understanding the factors that led to their dissatisfaction is essential to avoiding a recurrence of the same issues. Organizations could benefit from conducting exit interviews and maintaining open lines of communication with former employees to gain insight into their motivations for returning.

To prevent dissatisfaction that leads to turnover or boomerang behavior, organizations should focus on providing clear career development paths and growth opportunities. Ensuring that employees feel valued and perceive potential for advancement can help mitigate dissatisfaction, increase their commitment to the organization, and reduce turnover rates.

Organizations can develop targeted retention strategies by identifying the types of shocks that significantly affect employees' job satisfaction and their potential return. By proactively addressing issues related to career, work, and personal shocks, companies can reduce turnover and increase the likelihood of rehiring high-performing former employees (Lee and Mitchell 1994; Maertz and Campion 2004).

Establishing and maintaining strong alumni networks can facilitate informal job searches, making it easier for former employees to consider returning. Organizations can develop programs to keep former employees engaged, informed about company developments, and connected through professional networks (Barbulescu 2015; O'Connor 2013).

By understanding the mediating role of job dissatisfaction in the relationship between shocks and boomerang intention, managers can implement measures to monitor and improve job satisfaction continuously. This includes regular surveys, feedback mechanisms, and intervention programs aimed at mitigating negative affects and enhancing employee well-being (Brayfield and Rothe 1951; Spector 1997).

Companies can benefit from strategically rehiring boomerang employees who bring back valuable tacit knowledge and social capital. Developing structured reentry programs that facilitate seamless reintegration can maximize the contributions of returning employees and leverage their familiarity with the organizational culture and processes (Keller et al. 2021; Kogut and Zander 1992).

HR policies should incorporate provisions for supporting employees during and after shocks (career, work, or personal). This might include counseling services, flexible work arrangements, and career development opportunities that address the root causes of dissatisfaction and enhance overall job satisfaction (Holtom et al. 2005; Wilson and Schooler 1991).

Overall, this study contributes to the literature by offering a more integrated understanding of the pathways leading to boomerang intentions, emphasizing the significant roles of affective responses and job dissatisfaction.

Organizations can leverage these insights to develop more effective retention strategies and re-recruitment practices, ultimately enhancing employee stability and performance (Maertz and Campion 2004; Mitchell et al. 2001).

By recognizing the complex interplay between emotional reactions, job satisfaction, and informal job search behaviors, companies can better manage their human resources and foster a supportive and resilient workforce (Kogut and Zander 1992; Podolny and Page 1998).

By implementing these theoretical insights and practical strategies, organizations can better manage employee turnover, foster positive work environments, and harness the potential of boomerang employees to achieve sustained organizational success.

Appendix: Survey Items

Personal Questions

- Name
- Gender
- Age
- Position
- Career
- Career_total
- Tenure

Screening Questions

"Do you have personal experience of voluntary turnover?" (Yes, No)

Shocks: Career shock, work shock, personal shock

"Was there a particular jarring event that caused you to think about leaving? (Yes, No). Describe the shock in detail. Please characterize the shock as a career shock, a work shock, or a personal shock.

- Career shock: Project management issue, department and strategy issue, development issue, infrastructure issue, security work issue.
- Work shock: Job offer, learning opportunities, money, fight with boss or coworker, performance issue, merger and acquisition, layoff, irresponsible behavior.
- Personal shock: Spouse employment, illness, moved, school, own business, pregnant."

(Maier et al. 2021).

Shocks: Career shock, work shock, personal shock

Participants rated the degree to which each career shock impacted their career on a 5-point scale ranging from 0 (have not experienced the event; and thus, had no impact) to 4 (had a large impact).

"Was there a particular jarring event that caused you to think about leaving? (Yes, No). Describe the shock in detail. Please characterize the shock as a career shock, a work shock, or a personal shock.

- Career shock: Project management issue, department and strategy issue, development issue, infrastructure issue, security work issue.
- Work shock: Job offer, learning opportunities, money, fight with boss or coworker, performance issue, merger and acquisition, layoff, irresponsible behavior.
- Personal shock: Spouse employment, illness, moved, school, own business, pregnant."

(Maier et al. 2021).

Shocks: Career shock, work shock, personal shock

Participants rated the degree to which each career shock impacted their career on a 5-point scale ranging from 0 (have not experienced the event; and thus, had no impact) to 4 (had a large impact).

Shocks: Career shock, work shock, personal shock

- I just realized that I am pregnant.
- I just left my home city and we build a house in another city.
- My spouse got a job offer from another organization.
- My father got seriously ill.

Shocks: Career shock, work shock, personal shock

Participants rated the degree to which each career shock impacted their career on a 5-point scale ranging from 0 (have not experienced the event; and thus, had no impact) to 4 (had a large impact) (Seibert 2013).

Work shocks

- No further education opportunities to deal with new trends
- My current employer does not pay me the salary I want to have
- My colleague was performing better than me
- I had a clash with my supervisor due to organizational issues

(Holtom et al. 2005).

Career shocks

- A mentor or colleague who is important to you has left the organization.
- The organization experienced a significant negative event, such as a workforce reduction, bankruptcy, or major ethical scandal.
- You successfully took on a new task or tangible project that initially challenged you in a new and significant way.
- You received a pay raise, promotion, or desirable increase in responsibility sooner than expected.

(Seibert et al. 2013).

Negative Affect

Positive emotions: We measured positive emotions with three items that asked respondents to indicate how often they felt excited or enthusiastic during the job search during the preceding 3 months (using 5-point scales; α = .73). These items are indicators of the high activation dimension of positive affect and were adapted from the job affect scale (Burke et al. 1989). Negative emotions: Scared, distressed (Judge et al. 2006).

Job Dissatisfaction

We assessed job satisfaction using the five-item version of the Brayfield and Rothe (1951) measure. We measured all items using a 5-point Likert-type scale with 1 = strongly disagree and 5 = strongly agree. Participants indicated the extent to which they agreed with the following statements about their job each day: "At this very moment, I am unenthusiastic about my work" and "Right now, I feel fairly dissatisfied with my present job" (Judge et al. 2006).

Boomerang Intention

- I am willing to go back to my former company.
- Among many employment opportunities, I prefer to go back to my former company.

(Tian et al. 2022).

References

Adecco Group. 2023. *The Rise—and Return—of Boomerang Employees.* https://www.adeccogroup.com/future-of-work/latest-insights/the-rise-and-return-of-boomerang-employees

Barbulescu, Roxana. 2015. "The Strength of Many Kinds of Ties: Unpacking the Role of Social Contacts across Stages of the Job Search Process." *Organization Science* 26 (4): 1040–58.

Bentler, P.M., and Douglas G. Bonett. 1980. "Significance Tests and Goodness of Fit in the Analysis of Covariance Structures." *Psychological Bulletin* 88 (3): 588–606.

Bollen, Kenneth A., and Robert A. Stine. 1992. "Bootstrapping Goodness-of-Fit Measures in Structural Equation Models." *Sociological Methods & Research* 21 (2): 205–29.

Ferratt, Thomas W., Jayesh Prasad, and Harvey G. Enns. 2012. "Synergy and Its Limits in Managing Information Technology Professionals." *Information Systems Research* 23 (4): 1175–94.

Granovetter, Mark S. 1973. "The Strength of Weak Ties." *American Journal of Sociology* 78 (6): 1360–80.

Guimaraes, Tor, and Magid Igbaria. 1992. "Determinants of Turnover Intentions: Comparing IC and IS Personnel." *Information Systems Research* 3 (3): 273–303.

Hair, Joe, Carole L. Hollingsworth, Adriane B. Randolph, and Alain Yee Chong. 2017. "An Updated and Expanded Assessment of PLS-SEM in Information Systems Research." *Industrial Management & Data Systems* 117 (3): 442–58.

Hair, Joseph F., G. Thomas M. Hult, Christian M. Ringle, and Marko Sarstedt. 2021. *A Primer on Partial Least Squares Structural Equation Modeling (PLS-SEM).* 3rd ed. Sage.

Harman, Harry H. 1976. *Modern Factor Analysis.* University of Chicago Press.

Henseler, Jörg, Theo K. Dijkstra, Marko Sarstedt, Christian M. Ringle, Adamantios Diamantopoulos, Detmar W. Straub, David J. Ketchen, Joseph F. Hair, G. Tomas Hult, and Roger J. Calantone. 2014. "Common Beliefs and Reality about PLS." *Organizational Research Methods* 17 (2): 182–209.

Holtom, Brooks C., Terence R. Mitchell, Thomas W. Lee, and Marion B. Eberly. 2008. "Turnover and Retention Research: A Glance at the Past, a Closer Review of the Present, and a Venture into the Future." *Academy of Management Annals* 2 (1): 231–74.

Hom, Peter W., and Angelo J. Kinicki. 2001. "Toward a Greater Understanding of How Dissatisfaction Drives Employee Turnover." *Academy of Management Journal* 44 (5): 975–87.

ICAEW. 2023. *Boomerang Employees: Potential Risks and Rewards for Firms.* https://www.icaew.com/insights/viewpoints-on-the-news/2023/may-2023/boomerang-employees-potential-risks-and-rewards-for-firms

Judge, Timothy A., Brent A. Scott, and Remus Ilies. 2006. "Hostility, Job Attitudes, and Workplace Deviance: Test of a Multilevel Model." *Journal of Applied Psychology* 91 (1): 126–38.

Keller, JR, Rebecca R. Kehoe, Matthew Bidwell, David Collings, and Adam Myer. 2021. "In with the Old? Examining When Boomerang Employees Outperform New Hires." *Academy of Management Journal* 64 (6): 1654–84.

Kogut, Bruce, and Udo Zander. 1992. "Knowledge of the Firm, Combinative Capabilities, and the Replication of Technology." *Organization Science* 3 (3): 383–97.

Leary, Mark R., and Roy F. Baumeister. 2000. "The Nature and Function of Self-Esteem: Sociometer Theory." In *Advances in Experimental Social Psychology,* Vol. 32, pp. 1–62, edited by Mark P. Zanna, Academic Press.

Lee, Thomas W. 1988. "How Job Dissatisfaction Leads to Employee Turnover." *Journal of Business and Psychology* 2 (3): 263–71.

Lee, Thomas W., Terence R. Mitchell, Chris J. Sablynski, James P. Burton, and Brooks C. Holtom. 2004. "The Effects of Job Embeddedness on Organizational Citizenship, Job Performance, Volitional Absences, and Voluntary Turnover." *Academy of Management Journal* 47 (5): 711–22.

Lee, Thomas W., Terence R. Mitchell, Lowell Wise, and Steven Fireman. 1996. "An Unfolding Model of Voluntary Employee Turnover." *Academy of Management Journal* 39 (1): 5–36.

Lohmöller, Jan Bernd. 1989. *Latent Variable Path Modeling with Partial Least Squares.* Springer.

MacCrory, Frank, Vidyanand Choudhary, and Alain Pinsonneault. 2016. "Research Note—Designing Promotion Ladders to Mitigate Turnover of IT Professionals." *Information Systems Research* 27 (3): 648–60.

Maertz, Carl. P, and Michael A. Campion. 2004. "Profiles in Quitting: Integrating Process and Content Turnover Theory." *Academy of Management Journal* 47 (4): 566–82.

Maier, Christian, Sven Laumer, Damien Joseph, Jens Mattke, and Tim Weitzel. 2021. "Turnback Intention: An Analysis of the Drivers of IT Professionals' Intentions to Return to a Former Employer." *MIS Quarterly* 45 (4): 1777–1806.

Marin, Alexandra. 2012. "Don't Mention It: Why People Don't Share Job Information, When They Do, and Why It Matters." *Social Networks* 34 (2): 181–92.

Mitchell, Terence R., Brooks C. Holtom, Thomas W. Lee, Chris J. Sablynski, and Miriam Erez. 2001. "Why People Stay: Using Job Embeddedness to Predict Voluntary Turnover." *Academy of Management Journal* 44 (6): 1102–21.

Morgeson, Frederick P., and D. Scott DeRue. 2006. "Event Criticality, Urgency, and Duration: Understanding How Events Disrupt Teams and Influence Team Leader Intervention." *The Leadership Quarterly* 17 (3): 271–87.

Mossholder, Kevin W., Randall P. Settoon, and Stephanie C. Henagan. 2005. "A Relational Perspective on Turnover: Examining Structural, Attitudinal, and Behavioral Predictors." *Academy of Management Journal* 48 (4): 607–18.

Trimble O'Connor, Lindsey. 2013. "Ask and You Shall Receive: Social Network Contacts' Provision of Help during the Job Search." *Social Networks* 35 (4): 593–603.

Palich, L. 1995. "Managing in the International Context: Testing Cultural Generality of Sources of Commitment to Multinational Enterprises." *Journal of Management* 21 (4): 671–90.

Podsakoff, Philip M., and Dennis W. Organ. 1986. "Self-Reports in Organizational Research: Problems and Prospects." *Journal of Management* 12 (4): 531–44.

Porter, Caitlin M., Richard A. Posthuma, Carl P. Maertz, Janice R. Joplin, James Rigby, Mary Gordon, and Krisanne Graves. 2019. "On-the-Job and off-the-Job Embeddedness Differentially Influence Relationships between Informal Job Search and Turnover." *Journal of Applied Psychology* 104 (5): 678–89.

Rasmusen, Eric. 2006. *Games and Information: An Introduction to Game Theory*. Blackwell.

Shipp, Abbie J., Stacie Furst-Holloway, T. Brad Harris, and Benson Rosen. 2013. "Gone Today but Here Tomorrow: Extending the Unfolding Model of Turnover to Consider Boomerang Employees." *Personnel Psychology* 67 (2): 421–62.

SHRM. 2022. *Why You Should Welcome Back Boomerang Employees*. https://www.shrm.org/topics-tools/news/managing-smart/welcome-back-boomerang-employees

Spector, Paul E. 1997. *Job Satisfaction: Application, Assessment, Cause, and Consequences*. Sage.

Swider, Brian W., Joseph T. Liu, T. Brad Harris, and Richard G. Gardner. 2017. "Employees on the Rebound: Extending the Careers Literature to Include Boomerang Employment." *Journal of Applied Psychology* 102 (6): 890–909.

Trimble, Lindsey B., and Julie A. Kmec. 2011. "The Role of Social Networks in Getting a Job." *Sociology Compass* 5 (2): 165–78.

van Hooft, Edwin A., John D. Kammeyer-Mueller, Connie R. Wanberg, Ruth Kanfer, and Gokce Basbug. 2021. "Job Search and Employment Success: A Quantitative Review and Future Research Agenda." *Journal of Applied Psychology* 106 (5): 674–713.

Weiss, Howard M., and Russell Cropanzano. 1996. "Affective Events Theory: A Theoretical Discussion of the Structure, Causes, and Consequences of Affective Experiences At Work." In *Research in Organizational Behavior*, Vol. 18, pp. 1–74, edited by Barry M. Staw and Larry L. Cummings, Elsevier/JAI.

Zimmerman, Ryan D., and Todd C. Darnold. 2009. "The Impact of Job Performance on Employee Turnover Intentions and the Voluntary Turnover Process." *Personnel Review* 38 (2): 142–58.

Challenges and Choices for Digitizing a Labor Research Collection

CHARISSA O. JEFFERSON
ADRIANNA TARABOLETTI
Princeton University

To enhance discoverability and wide public access, Princeton University's Library and Industrial Relations Section has embarked on a large-scale digitization project of its finite collection of over 31,000 original items produced by companies, trade unions, research organizations, governmental agencies, and professional associations spanning from 1922 through 1989. This collection is significant to the labor research community as the holdings prove unique to the study of 20th-century labor history and modern national economic policy development. The poster highlights the variety of materials and takeaways from a 100-box pilot project in developing the structures and mechanisms for the lifecycle of the project.

The Impact of Wage Negotiation Opportunities on Job Performance: A Focus on Self-Efficacy

GIWOO KIL
Korea University

Wages serve as a powerful motivator for enhancing job performance. However, there is a scarcity of research on wage negotiations in the context of South Korea, largely due to the prevalent use of fixed salary structures, limiting the opportunities for wage negotiations. This poster explains whether the mere presence of wage negotiation opportunities has a positive impact on job performance. Additionally, it examines whether individuals with high self-efficacy, regardless of the negotiation outcomes, experience an improvement in job performance through wage negotiations. To achieve this, data will be collected from 100 full-time employees representing various job sectors. Hierarchical regression analysis will be conducted for hypothesis testing. Enhancing employee morale and engagement requires the inclusion of employees in decision-making processes. This research intends to contribute to activating negotiation practices within organizations, ultimately enhancing employee performance and participation.

Museum of Employment in Sabah (Malaysia): Idea and Reality from Different Perspectives in the Social Sciences

MAHADIRIN BIN HJ. AHMAD
KEE Y. SABARIAH KEE MOHD YUSSOF
University of Malaysia Sabah

SHARIFAH NURSYAHIDAH SYED ANNUAR
University Kebangsan Malaysia

BASZLEY BEE BASRA BEE
NORHUDAH SALLEH
University of Malaysia Sabah

Museums serve as knowledge sources, focusing on past events and experiences. Therefore, we want to focus on a learning space for society, particularly in industrial and employment relations. The case study of Sabah Forest Industries (SFI), which used to be the only pulp and paper mill in Malaysia, illustrates this concept. Over three decades, operations need to stop, impacting workers and causing conflict. Workers struggle to secure recognition for Sabah Forest Industries Employee Union (SFIEU). Management practices led to union busting, resulting in SFIEU affiliating with the national union Sabah Timber Industries Employee Union (STIEU). Thus, this poster presents the idea of restoring SFI's structure as a source of employment for Sabahan and foreign labor in the Employment Museum. The ideas are explained from different perspectives in social sciences—namely, from collective action, political science, historical, media, and dispute resolution.

High-Performance Work Systems and Profit: Positive Correlation and Prescriptive Truthiness

YINGCHUN WANG
University of Houston Downtown

Abstract

Empirical evidence often shows a positive correlation between high-performance work systems (HPWS) and firm performance, leading to widespread advocacy for HPWS adoption. However, our theoretical model suggests that this correlation is driven by contextual variables, such as firm-specific production technology and market conditions. Our findings indicate that while HPWS can enhance operational performance, it may worsen financial outcomes if misaligned with a firm's unique context. These insights challenge the universal applicability of HPWS, urging caution in policy recommendations based on empirical correlations alone.

Introduction

High-performance work systems (HPWS) are widely promoted in human resources management education and practice as a means to boost employee commitment, skills, and collaboration, leading to enhanced organizational performance. This advocacy is supported by numerous empirical studies reporting a positive correlation between HPWS and firm performance. However, our research challenges the universal applicability of HPWS, highlighting the nuanced reality that not all firms benefit equally. By employing a formal model rooted in contingency theory, we demonstrate that the effectiveness of HPWS depends heavily on firm-specific factors, such as production technology and market conditions. This paper is the first to formalize the relationship between HPWS and firm performance from a contingency perspective, offering a more cautious interpretation of the widely observed positive correlation. Our findings have important implications for HRM practitioners, urging careful consideration of context before adopting HPWS.

HPWS and Output Value

We model the relationship between HPWS and operational performance, specifically output value, using the Cobb–Douglas production function. This widely used production function has been used to capture the diverse inputs relevant to HPWS, such as knowledge and human capital. By establishing this foundational link, we set the stage for examining how HPWS interacts with other inputs and contextual factors to influence profit, thereby revealing the conditions under which HPWS can be the most beneficial.

Results of the Model: Exogenous Variables and the Effect of HPWS on Output Value

All the exogenous factors influence the effect of HPWS on output value. First, a higher HPWS output elasticity can amplify the effect of HPWS. The output elasticity usually correlates with the demand for human capital, which is determined by a firm's production technology. This implies that firms reliant on a higher level of human capital, such as technology adopters or customized service providers, will experience a more pronounced effect from HPWS. In contrast, standardized production, requiring a lower level of human capital, leads to a smaller output elasticity of HPWS, thus diminishing HPWS's impact.

Factors not directly related to HPWS but boosting baseline output value also enhance HPWS's effect on output value since these factors and HPWS work in tandem to magnify each other's impact on output value. For instance, when there's an increase in capital investment, the effects of HPWS will be heightened. An increase in capital investment can be due to a rise in capital output elasticity or a decrease in unit financing costs, so these two changes can also magnify the effects of HPWS. Similarly, greater labor-hour input elevates HPWS's impact. Hence, an increase in labor output elasticity or a decrease in labor hour price not only increases the optimal labor-hour input but also subsequently amplifies HPWS's influence on the output value.

HPWS and Profit in Optimization

After modeling the effect of HPWS on output value, we focus on financial performance, specifically profit, by including the unit costs of HPWS investment, capital, and labor in the model to examine the impact of exogenous variables when HPWS and profit are optimized.

Results of the Model: Exogenous Variables and HPWS Investment

The optimal level of HPWS investment increases with the output elasticity of HPWS. This aligns with the empirical findings. The output elasticity reflects the demand for human capital. Sanchez et al. (1999) found a positive correlation between high-involvement HR practices and the adoption of new technologies and shifts in business environments requiring superior human capital. Rodwell and Teo (2008) reported a positive relationship between human capital enhancing HR practices and customer demands for quality, service, and technical specifications. Pil and Macduffie (1996) also observed a connection between a plant's new product proportion and the later stages of high-involvement work practice adoption.

The optimal level of HPWS investment decreases with the HPWS cost. Empirical evidence appears to support this relationship. The costs of selection, training, and motivation may be mitigated by a higher average quality of human capital in the labor market. Lawler et al. (2011) found a positive correlation between HPWS implementation in US firms' foreign branches and the foreign country's education spending per capita. HPWS costs may also be reduced by leveraging internal communication technology for efficient participatory decision-making and by training younger employees, who may potentially be more adaptable and trainable. Kauhanen (2009) found higher HPWS participation with the use of communication technology or among younger employees.

Unlike the above exogenous variables, which directly relate to HPWS, other variables only indirectly affect optimal HPWS investment through the output value. This aligns with the findings of Shin and Konrad (2017) that higher past productivity led to increased HPWS adoption later on.

Higher product prices may result from limited competition, which could explain why, as Wood and de Menezes (1998) found, UK firms with less competition more often adopted HPWSs. Total factor productivity, being the primary source of growth rate (Wang et al. 2021), may be proxied by the growth rate. HPWS implementation was indeed found to be positively correlated with a host country's growth rate (Lawler et al. 2011). Total factor productivity also varies by industry. For instance, the textile industry trails other sectors in terms of both total factor productivity and its HPWS index (Blasi and Kruse 2006).

Since the optimal output value increases with optimal capital input, the optimal HPWS investment also rises with capital input. This is due to the complementarity between HPWS and capital. Phuong's study (2020) found a positive link between registered capital and the degree to which the practices of HPWS were implemented.

As capital input increases with output elasticity of capital and drops with unit financing costs, HPWS investment is similarly affected. Consistent with these relations, Datta et al. (2005) observed that higher capital intensity, a proxy for the output elasticity of capital in competitive markets, is linked to more extensive HPWS coverage among employees. Chow et al. (2008) found that Chinese collectively owned firms, which tend to have more loan challenges and hence higher capital costs, use fewer high-commitment HR practices.

Similarly, HPWS and labor input are complementary, as increased labor hours amplify the effect of HPWS investment. Kroon et al. (2013) found a positive relationship between HPWS practices and employment size. Since labor hour input increases with output elasticity of labor and decreases with unit labor hour cost, HPWS investment also increases with output elasticity of labor and decreases with labor hour cost. Consistent with these expectations, Lawler et al. (2011) found that less restrictive foreign labor laws, implying lower labor costs, increased HPWS implementation in US firms' foreign branches. Yet few HPWS studies directly investigate the output elasticity of labor hours.

Exogenous Variables and Profit

The discussed exogenous variables affect firm profit in the same direction as they affect HPWS. Conceptually, this is because if they increase (decrease) the optimal profit, it implies that they also enhance (diminish) the optimal output value. This, in turn, amplifies (reduces) the marginal output value of HPWS investment and hence increases (decreases) the optimal HPWS investment correspondingly.

Implications of the Model

The model predicts a positive relationship between HPWS and profit, particularly in competitive environments where optimizing inputs is essential. However, this relationship may be weaker in nonprofit organizations or firms with goals beyond profit, such as employee welfare or social impact.

While the model does not examine other financial performance metrics, it suggests that the same contextual variables affecting HPWS and profit are likely to influence other metrics, such as return on assets and stock market performance, in a similar direction.

Studies that fail to control for firm heterogeneity may find misleading positive correlations between HPWS and performance. The extent of these spurious effects depends on the variation and interrelationships of confounding factors, which can either intensify or mitigate cross-sectional relationships. For example, firms within a single industry with similar production technologies may show weaker correlations between HPWS and profitability compared to more diverse samples, as the variations in the confounding factors are limited.

Finally, the model has implications for longitudinal studies that control for unobserved firm heterogeneity. The relationship between HPWS and financial performance within firms may be ambiguous, depending on whether there is under- or overinvestment in HPWS. Positive relationships might indicate an underinvestment in HPWS, while negative relationships could suggest overinvestment. However, a positive relationship between HPWS and financial performance may also result from changes in confounding variables over time.

Conclusion

While our analysis primarily focuses on profit and financial performance, the same confounders (except HPWS cost) can also inflate the relationship between HPWS and output value, which relates to operational performance.

This study contributes to three key areas. First, it provides the first formal model explaining the positive link between HPWS and firm performance from a contingency perspective, highlighting the significant difference between within-firm and across-firm relationships. We identify specific potential confounders, some overlooked in the literature, and suggest that *any* contextual variables influencing a firm's optimal output value may similarly affect HPWS investment. This leads to the second contribution: implications for empirical studies. Contextual variables arising from firm heterogeneity can introduce omitted variable bias, which is often underexplored in HPWS research. Researchers should control for unobserved heterogeneity using methods such as fixed effects or first-difference estimators, although each has its limitations. Finally, we caution practitioners against adopting HPWS based solely on cross-firm correlations and emphasize the need for context-specific HR practices. This contingency framework can apply to any nonessential HR practices with nondecreasing marginal costs, not just HPWS.

We conclude by proposing future research directions to address the model's limitations, such as exploring the impact of uncertainty and examining how relaxing the model's assumptions affects optimal HPWS investment. The relationship between HPWS and firm performance, despite extensive study, remains complex and warrants further investigation.

References

Blasi, J.R., and Kruse, D.L. (2006). U.S. High-Performance Work Practices at Century's End. *Industrial Relations: A Journal of Economy and Society 45* (4), 547–578. https://doi.org/10.1111/j.1468-232X.2006.00443.x

Chow, I.H., Huang, J.-C., and Liu, S. (2008). Strategic HRM in China: Configurations and Competitive Advantage. *Human Resource Management 47* (4), 687–706. https://doi.org/10.1002/hrm.20240

Datta, D.K., Guthrie, J.P., and Wright, P.M. (2005). Human Resource Management and Labor Productivity: Does Industry Matter? *The Academy of Management Journal 48* (1), 135–145. https://doi.org/10.2307/20159645

Kauhanen, A. (2009). The Incidence of High-Performance Work Systems: Evidence from a Nationally Representative Employee Survey. *Economic and Industrial Democracy 30* (3), 454–480. https://doi.org/10.1177/0143831x09336560

Kroon, B., Van De Voorde, K., and Timmers, J. (2013). High Performance Work Practices in Small Firms: A Resource-Poverty and Strategic Decision-Making Perspective. *Small Business Economics 41* (1), 71–91.

Lawler, J.J., Chen, S., Wu, P.-C., Bae, J., and Bai, B. (2011). High-Performance Work Systems in Foreign Subsidiaries of American Multinationals: An Institutional Model. *Journal of International Business Studies 42* (2), 202–220.

Phuong, T.H. (2020). Impact of Organizational Characteristics, CEO Education, and Firm Ownership on the Adoption and Effectiveness of High Performance Work System in Vietnam. *Organizations and Markets in Emerging Economies 11* (21), 83–106.

Pil, F. K., and Macduffie, J. P. (1996). The Adoption of High-Involvement Work Practices. *Industrial Relations 35* (3), 423–455. https://doi.org/10.1111/j.1468-232X.1996.tb00414

Rodwell, J. J., and Teo, S. T. T. (2008). The Influence of Strategic HRM and Sector on Perceived Performance in Health Services Organizations. *The International Journal of Human Resource Management 19* (10), 1825–1841. https://doi.org/10.1080/09585190802323934

Sanchez, J. I., Kraus, E., White, S., and Williams, M. (1999). Adopting High-Involvement Human Resource Practices: The Mediating Role of Benchmarking. *Group & Organization Management 24* (4), 461–478. https://doi.org/10.1177/1059601199244004

Shin, D., and Konrad, A. M. (2017). Causality Between High-Performance Work Systems and Organizational Performance. *Journal of Management 43* (4), 973–997.

Wang, Y., Ye, G., Zheng, C., and Zhang, S. (2021). Total Factor Productivity Growth and Its Decomposition in the Chinese Construction Industry Since China's Accession to the World Trade Organization. *Engineering, Construction and Architectural Management* (ahead of print). https://doi.org/10.1108/ECAM-09-2020-0741

Wood, S., and de Menezes, L. (1998). High Commitment Management in the U.K.: Evidence from the Workplace Industrial Relations Survey, and Employers' Manpower and Skills Practices Survey. *Human Relations 51* (4), 485–515. https://doi.org/10.1177/001872679805100403

Algorithmic Management: Safeguarding Workers' Rights in Automated Decision-Making

SONIA BASTIGKEIT ERICSTAM

Stockholm University

Introduction

Workers are particularly vulnerable to risks associated with the use of artificial intelligence (AI),[1] given the imbalance of power present in the employment relationship.[2] In the context of AI's use in employment, the application of automated algorithms and automated decision-making (ADM) systems in managerial functions—commonly referred to as algorithmic management—has emerged as a key issue.

In the European Union (EU), the last months have seen important developments in the legal protection of workers subject to such management. The adoption of the AI Act[3] (AIA), which establishes rules on the marketing, deployment, and use of AI systems, alongside the Platform Directive[4] (PFD) which introduces sector-specific regulations on algorithmic management in the gig economy, has formally strengthened the rights of workers under algorithmic management.

In addition, given that the AIA does not impact data protection in the General Data Protection Regulation[5] (GDPR), the processing of individual data is central to the functioning of algorithms, and automation of managerial decisions is the basis for algorithmic management. The prohibition of ADM in Article 22 GDPR means that the use of this form of management is, in principle, prohibited.

The purpose of this paper is not to assess to what extent the legal framework governing the use of algorithmic management caters to the protection of certain specific material rights of workers (such as privacy rights or protection from discrimination). Rather, I analyze the strength of the legal framework applicable to this type of management in providing workers and their representatives with a legal infrastructure by which to claim workers' rights. In this regard, the extent to which the correct use of ADM systems can be secured ex ante, the extent to which automated managerial decisions can be understood, and the extent to which accountability can be sought are central. Furthermore, I analyze the rules applicable to this type of management as a new set of limitations on the managerial prerogative in that the rules restrict the ways in which the employer can execute their managerial functions. As such, the legal framework has the potential to challenge our concept of the dynamic between the employer and workers in a setting where management is automated.

The European Union (EU) provides an example of a legislative strategy by which the protection of the rights of workers subject to algorithmic management is found in various legal acts. Although the main purpose of the GDPR is the protection of individual data, the AIA mainly includes obligations that different actors responsible for AI systems need to fulfill and report to authorities. The PFD includes detailed rules to protect people performing work in the platform sector, governing what platform companies need to do in relation both to supervising authorities as well as to individuals performing work and their representatives.

Given that neither the PFD nor the AIA has yet been implemented, it is not possible to evaluate their practical effects on the protection of workers' rights in the algorithmic management context. However, even in this early stage of this legislation, it is possible to theoretically analyze the potential effects of the legal framework and to point to areas of interest both for the practitioners and courts who will implement them in the not very distant future.

Algorithmic Management and Automated Decision-Making

Background

An algorithm is a set of instructions that carry out commands.[6] Algorithms can also be described as processes or sets of rules that can be followed in calculating or other forms of problem-solving.[7] The first time the term "algorithmic management" appears to have been used in the employment context was in 2015. It was then defined as the use of "software algorithms that assume managerial functions and surrounding institutional devices that support algorithms in practice."[8] Nguyen and Mateescu (2019) described algorithmic management as the use of technological tools and techniques to manage workers remotely, where collection of workers' data to enable automated or semi-automated decision-making is a central aspect.[9]

The use of algorithms in management functions is a continuance of the use of technological tools applied by managers to organize work. As such, algorithmic management consists of aspects that have historically been fundamental in the employment relation, such as surveillance and control of the work performed, but also of aspects that could be defined as new in the sense that technical systems are used not only as tools but have a more qualitative impact in their application. To distinguish the use of algorithmic management from previous forms of technology use in management, the concrete functioning of the algorithms in that they can be used to automate managerial decisions is central. For the data to be used for automated management decisions, the employer must make use of tools to process and analyze what has been captured in the data. This is enabled by machine learning, where probabilistic analyses of large datasets can be implemented and where sophisticated statistical modeling is used to recognize patterns or correlations in the data.[10] Such datasets can be collected because employers can track their employees over time through devices such as smartphones, desktops, tablets, vehicles, and wearables.[11]

Algorithmic management can be divided into three forms of automated managerial functions: direction, evaluation, and discipline of workers. Direction involves deciding what sort of work should be performed and when it should be performed. Evaluation entails reviewing the work performed by workers to assess their performance and identify individuals who do not meet the standards. Discipline involves decisions about reprimands and rewards of workers.[12]

The GDPR

All use of data in an algorithmic management context must be based on lawful grounds, per Article 6 of the GDPR.[13] Accordingly, the central tenet for algorithmic management in the GDPR is Article 22, which provides that a data subject not to be subject to a decision based *solely* on automated processing. In addition to being solely based on automated processing, for the decision to be prohibited, the effects of the decision need to be *legal* or affect the data subject in a *similarly significant* way. In December 2023, in the ruling on SCHUFA Holding (Scoring),[14] the Court of Justice of the European Union (CJEU) provided some guidance on how the criteria in the article are to be interpreted. The CJEU clarified that, notwithstanding that the wording of the article suggests it includes a right that needs to be invoked by the data subject, the article does in fact constitute a prohibition.

Furthermore, the CJEU emphasized that in a situation where an automatically established value based on data relating to an individual is transmitted from a data controller to a third-party controller, and the latter, in accordance with consistent practice, draws strongly on the automatically established value in making a decision, the decision is to be considered solely automated, even if a human formally makes the ultimate decision (in this case, the human was an employee at a credit institute who had to decide whether to grant an individual a loan on the basis on an automated credit score transferred from a credit information agency).[15] The CJEU noted that a restrictive interpretation that would consider the automated credit score a preparatory act and not a decision would risk creating a lacuna in legal protection for the data subject. According to the CJEU, such interpretation would entail that the prohibition in Article 22 of the GDPR could be circumvented.[16] Furthermore, the CJEU stated that unless the crucial point for the decision is placed where the automated credit score is generated, the data subject would not be able to assert her information rights in accordance with Article 15(1)(h) of the GDPR, which regulates information rights of the data subject in ADM because it was the credit information agency and not the credit institute who had that information.[17]

When applying the reasoning of the CJEU in the context of algorithmic management, one can imagine a situation in which a human manager is choosing which employees to promote. In the selection process, the manager uses an automated system in which an algorithm sorts employees based on performance scores. The manager will, as the basis for her promotion decision, be given a ranking of employees suggested for promotion (Group A) and employees who are not considered deserving of promotion (Group B). If we imagine that the manager has historically had no reason to question the validity of the algorithmic performance evaluation tools used, it can at least be questioned whether the decision of the manager to carry through the promotion of employees sorted in Group A will be anything more than a mere "rubber stamping" operation, where it is questionable whether her human involvement is qualitative. Depending on where the critical point for the decision is placed (in the assortment made by the algorithm or when the manager makes the decision based on the algorithmically produced proposal) and depending on to what extent one considers the human involvement qualitative, the decision could be deemed to be either solely based on automated processing or not solely based on automated processing. Consequently, the decision can be deemed allowed or prohibited in accordance with Article 22 of the GDPR. In my opinion, the situation would be considered as being based solely on automated processing, even though it does not entail the transfer of an automated score from a third party, and also as having such severe effects on the data subject that the decision would meet the criterion for being prohibited.

If the decision is considered to be based solely on automated processing and if the effects of the decision are decided to be legal or of similar significance, the decision might still be allowed under Article 22. This follows from the fact that the prohibition in the article has three exceptions. The first exception is that the ADM can be based on necessity in order to enter into or to perform in accordance with a contract between the data subject and a data controller. The second exception is that an ADM that would otherwise have been prohibited in accordance with the article may be authorized by law in the Member State or EU if such law includes suitable measures to safeguard the rights, freedoms, and legitimate interests of the data subject. The third exception is that the ADM is legal if it is based on the explicit consent of the data subject. In relation to the exceptions based on contractual necessity and consent, the data controller should implement certain suitable measures to safeguard the rights, freedoms, and legitimate interests of the data controller, in that the data subject must at least have the right to obtain human intervention on the part of the data controller, to express her point of view, and to contest the decision.

To assess the scope of the exception based on contractual necessity, the guidelines in Article 29 Working Party[18] (the advisory body for the interpretation of the GDPR) on ADM can be considered. According to the advisory body, necessity should be given a narrow interpretation. The data controller must be able to show that to achieve the purpose of the processing of personal data underlying the ADM, no other effective and less privacy-intrusive means are available.[19] Examples of reasons that are not sufficient to motivate the use of ADM are the aim for greater consistency, fairness, or improved efficiency in the decision-making process.[20] The processing of salary information and bank account details so that wages can be paid is an example of the performance of the contract data. However, the example is not specifically related to the context of Article 22 of the GDPR (somewhat confusingly, both consent and contractual necessity are also examples of lawful grounds for processing of data in Article 6 of the GDPR). As an example of when ADM may be allowed on the grounds of this exception, a situation where the use of automated systems is necessary in recruitment processes because of the amount of applications is mentioned.[21] The connection between the assessment of necessity on the one hand and compliance with the fundamental principle of purpose limitation in GDPR on the other hand, entailing that personal data should be collected only for specified, explicit, and legitimate purposes and not further processed in a manner incompatible with those purposes, is also mentioned in the Article 29 Working Party. In this regard, it is important to determine the exact rationale of the contract—in other words, its substance and fundamental objective—as it is against those that the test on whether the data processing is necessary for the performance must be pursued.

According to Article 29 Working Party, the exception based on consent is not suitable in the algorithmic management context, given the imbalance of power present between the employer and the worker. The dependency resulting from this imbalance makes it unlikely that the employee can withdraw the consent without risking negative effects.[22] Furthermore, in accordance with the GDPR, consent to data processing must be able to be withdrawn at any moment, which for practical reasons makes it an unsuitable ground in the algorithmic management context.[23] This leaves us with the conclusion that the only viable exception allowing the use of ADM in the employment context is the one based on Member State or union law, provided that such law meets the criterion of providing suitable measures as safeguards.[24] In the SCHUFA Holding (Scoring) ruling, the CJEU stated that when ADM is allowed in accordance with law, such measures should entail an obligation for the data controller to use adequate and qualitative procedures for establishing that the risk for mistakes in the systems used are minimized and making sure that potential risks for the interests and rights of the data subject, including the risk of discrimination, are prevented. According to the CJEU, such measures require at a minimum that the data subject must have the right to personal contact with the data controller in order to be able to express their opinion and to contest the decision.[25]

Regarding the adoption of Member State law on the use of ADM for management purposes, Article 88 of the GDPR is relevant. This article entails that Member States may, through the use of law or collective agreements, provide for more-specific rules to ensure the protection of the rights and freedoms in respect to the processing of employee personal data. Article 88 of the GDPR stipulates that such rules must meet certain standards. The rules must (1) provide more-specific regulations to ensure the protection of the rights and freedoms in respect to the processing of employee's personal data and (2) include suitable and specific measures to safeguard the data subject's human dignity, legitimate interest, and fundamental rights. In March 2023, in the ruling *Hauptpersonalrat der Lehrerinnen und Lehrer beim Hessischen Kultusmnisterium v. Minister des Hessischen Kultusministerium*, the CJEU clarified that in order for Member State law to be allowed under Article 88 of the GDPR, it must meet the requirements of both (1) and (2).[26] According to the CJEU, the Member State law on data processing in the employment context cannot be considered more specific if it does not meet the criterion of ensuring the protection of workers' data rights and includes suitable and specific measures mentioned in the article.

The AIA

The AIA reflects the focus on the decision-making aspect of the automated systems used in managerial functions, as the definition of such AI systems is that they are "intended to be used to make decisions affecting terms of work-related relationships, the promotion or termination of work-related contractual relationships, to allocate tasks based on individual behavior or personal traits or characteristics, or to monitor and evaluate the performance and behavior of persons in such relationships."[27] The general definition of AI systems in the AIA is a "machine-based system that is designed to operate with varying levels of autonomy and that may exhibit adaptiveness after deployment, and that, for explicit or implicit objectives, infers, from the input it receives how to generate outputs such as predictions, content, recommendations that can influence physical or virtual environments."[28] The AIA definition of AI is broad but, as it seems, the crucial point for a system to qualify as an AI system in the AIA is that the system *infers* output. How the concept of inference is to be interpreted is elaborated on in the recitals of the AIA.[29] In sum, in order to be encompassed by the AIA high-risk definition when used in managerial functions, the systems must (1) be able to infer output from input and (2) have the intended use to make decisions in the management context. In the recitals to the AIA, it is stated that the AIA does not impact any rights of data subjects in data protection law, and in this regard, rights relating to automated decision-making are particularly mentioned.[30]

The PFD

By next year, approximately 43 million people will perform work through platform companies within the EU.[31] The PFD aims to strengthen the legal protection of these workers.[32] In the PFD, algorithmic management functions are described as "automated monitoring and decision-making systems powered by algorithms" that replace functions that managers usually perform in businesses. Among such functions, allocation of tasks, pricing of individual assignments, determining work schedules, giving instructions, evaluating the work performed, and provision of sanctions or incentives are mentioned.[33] The PFD includes rules on automated systems used for surveillance and management purposes.[34] In the PFD, the rights of individual data subjects in relation to ADM in the GDPR are acknowledged. However, the PFD states that in the context of algorithmic management in platform work, considering the serious impact on people performing such work, certain decisions with particularly grave consequences for those performing the work should always be made by a human.[35] The PFD also includes more specific rules in relation to the GDPR. The PFD defines ADM systems as systems "used to [m]ake or support, through electronic means, decisions that significantly affect persons performing platform work." [36] In relation to the rules on ADM in the GDPR, the PFD includes limitations on what type of information can be processed in automated managerial decisions and, in relation to the GDPR, it includes additional transparency obligations that need to be met when those systems are used.

The Context: The Managerial Prerogative

To understand why the rules governing ADM are relevant for the employment relationship dynamic, the fundamentals of management and the characteristics of the employment contract must be contextualized. In 1937, the British economist Ronald Coase explored the grounds for the decision to operate in the form of a firm, in contrast to other forms of operations. He concluded that, through the formation of a firm, some of the costs of being on the market are avoided. As an example, the contract between the firm and the employee replaces several contracts that would otherwise have had to be completed.[37] In echoing the ideas articulated by Coase, Elizabeth Anderson, 70 years later, pointed to the benefits of operating in a firm such that the hierarchical structure, where the employee is subordinate to the manager, leads to efficiency because the managerial authority is open-ended and can encompass the unknown circumstances that may appear in the context in which the work is to be carried out.[38]

The managerial prerogative is considered a general principle of law, and the material content of this right encompasses the organization of work, production, methods of work, and working hours.[39] The managerial prerogative is, in the United States, just as in European countries, considered a rule with the status of being the default norm in the workplace. This means that, in the absence of contractual agreements, law, or case law, every decision in the workplace is taken within the scope of the discretion of the employer.[40] In other words, when an employment contract is agreed on between the employer and the employee, the managerial prerogative provides that the employee is subject to direction, supervision, and discipline that the employee has to obey without the employer having to obtain the consent of the employee.[41]

The managerial prerogative can be described as a residual competence, constituting the sphere in which employment relation matters are not covered by law, collective agreements, or individual contracts. It is in this sphere that the manager has discretion to make unilateral decisions.[42] Limitations of the managerial prerogative have, in a Nordic context, been analyzed in light of three competences: the material, the formal, and the personal competence. Accordingly, collective agreements or laws, including provisions for salary levels, working time, or other conditions of work, constitute examples of limitations of *material* competence. This is because those provisions require that a manager cannot freely decide which decisions to make or which content those decisions should have in relation to the workers.[43] On the other hand, provisions in collective agreements or laws regarding requirements for information and consultation and for how material decisions should be made constitute limitations of the *procedural* competence of the employer. This is because those limitations affect the scope of discretion about how a manager should make decisions.

However, the *personal* competence of the manager is not limited because it is the manager who makes the final decision. [44]

The duty to inform and consult employees, which can be considered a *procedural* limitation of the managerial prerogative, can also be considered as accentuating the managerial prerogative, given that, notwithstanding the inclusion of the collective of workers in the process, the final decision is still made by the employer.[45] In accordance with the duty of the employer to negotiate matters with the collective of workers in certain situations, the employer has to present the grounds for certain managerial decisions. These grounds can, in principle, not be subject to legal review, but the duty to present the grounds serves the purpose of counteracting unreasonable managerial decisions. The extent to which managerial decisions can undergo legal review is limited to where those decisions conflict with the law. The extent of the managerial prerogative can be described as the boundaries for what type of decisions the employer has a legal right to decide unilaterally.[46] This means, in principle, that workers have few possibilities to claim that managerial decisions are not fair.[47] As long as the managerial decisions are taken within the ambit of the managerial prerogative, the manager has no obligation to motivate the decision. The part of the managerial prerogative entailing the right to lead the work can be described as constituting the last strong bastion of the managerial prerogative.[48]

Preconditions for Safeguarding Workers' Rights in Algorithmic Management

Ex Ante Protection

The fundamental principle of the managerial prerogative and examples of the limitations of the prerogative were described above. The legal framework governing the introduction and use of algorithmic management entails rules aimed at preventing harm to workers' rights to emerge. Among those rules, Article 22 of the GDPR is central, given that it is a prohibition and not a right that needs to be invoked by the worker in the capacity of the data subject. In the context of algorithmic management, provisions in the relevant legal framework entailing the obligations of the employer who wants to introduce and use algorithmic management systems can be seen as limitations on the managerial prerogative.

In the AIA, there are certain requirements that need to be fulfilled by the deployer of AI systems. In the context of algorithmic management, the employer falls under the definition of the deployer.[49] The responsibilities of the deployer of AI systems are found in Article 26 of the AIA. Article 26(9) of the AIA stipulates that the deployer should use the information provided under Article 13 of the AIA to pursue an impact assessment when such assessment is required under Article 35 of the GDPR. Such impact assessment should include at least a systematic description of the planned processing of data and its purposes, an assessment of the need and the proportionality of the processing in relation to its purposes, an assessment of the risks for the rights and freedoms of the registered persons, and the actions planned to handle the risks, including protective actions and routines to ascertain the protection of the personal data. Furthermore, following Article 26(7) of the AIA, the deployer needs to inform affected workers and their representatives before the introduction of AI systems.

In accordance with the AIA, the use of AI systems in managerial functions is considered a form of high-risk use.[50] Hence, certain requirements need to be met before the introduction of such systems to the market. The responsibility for fulfilling these requirements falls on the provider of the AI systems.[51] These requirements stipulate that AI systems must have a risk management system; that appropriate data governance and practices for training, validation, and testing of models and datasets are in place; that sufficient technical documentation is available; that record-keeping is sufficient; and that transparent system design and instructions for use are ensured. Furthermore, there is a requirement for human oversight. The aim of human oversight is to improve transparency in the operations of the system. However, in the case of AI systems applied in the employment context, whether the requirements are met is subject to self-assessment by the provider of the AI system.[52] That these requirements should be met before the introduction of the systems can be considered a means to prevent harm from occurring—in other words, expressing an ex ante approach to risks associated with the use of high-risk AI systems.

Under certain circumstances, the responsibilities of the provider of the AI system are taken on by the deployer of the AI system. These circumstances are listed in Article 25 of AIA. Among the situations in which this happens is if substantial changes are made to the AI system or if the purpose of the use of the AI system is changed. In the preamble to the AIA, the meaning of the concept of substantial change is elaborated upon. There, it is established that when a high-risk AI system has undergone changes that may affect compliance with the AIA or when the system has undergone changes in accordance with the established term substantial change (referring to products regulated through the harmonization rules in the EU), they should undergo a new conformity assessment. A distinction is made between such cases and situations that should not constitute substantial changes. The latter are exemplified by changes of the algorithm and the AI system that continues its learning after they have been introduced on the market or been used (e.g., automatically adjusts how the functions are exercised) if such changes have been established beforehand by the provider and have been assessed when the conformity assessment was made.

The PFD includes several rules aimed at preventing risks from materializing. Article 8 includes an obligation to pursue a risk assessment in accordance with Article 35 of the GDPR. Following Article 8 of the PFD, the platform company should seek the views of people performing platform work and their representatives. Article 9 of the PFD includes rules on information for platform workers and their representatives before the introduction of automated managerial systems. Articles 13 and 14 of the PFD suggest that EU directives on information and consultation should not be affected by the PFD. However, it is stated that Member States of the EU should make sure that workers' representatives or workers are informed and consulted on decisions likely to lead to the introduction of, or to, substantial changes in the use of algorithmic management systems.

Explainability

Algorithmic programs are sometimes described as "black boxes," meaning that they will make predictions and decisions in a way similar to humans but without being able to communicate the grounds for the outcome of the decision-making process.[53] The individual who is subject to an automated decision rarely has the technical knowledge necessary to understand why the result was generated from the inputs and might also be unaware of what inputs are used in the algorithm.

The contents of the information rights in the GDPR regarding ADM have been debated. On the one hand, the fact that the right to an explanation of an automated decision, found in the preamble, did not make it to a legally binding article has been suggested to show that the GDPR does not entail a right to explanation.[54] Furthermore, the wording of the rights in Articles 13, 14, and 15 of the GDPR has been assessed as including only a right to information ex ante, not a right to explanation ex post of automated decisions.[55] Contrasting rights, such as the property rights of the owner of the data processing systems, have been defined as obstacles for the data subject to obtaining information regarding the functioning of the automated systems used in the data processing.[56] On the other hand, it has been argued that the right to contestation found in Article 22 of the GDPR must implicitly entail a right to explanation of the decision because the appeal right would be hollow without it.[57]

In SCHUFA Holding (Scoring), the CJEU did not elaborate on the extent of information rights under Article 15 of the GDPR in relation to ADM. However, in the opinion of Advocate General Pikamäe, delivered in the case before the judgment, some guidance is provided on the scope of those information rights.[58] According to the Advocate General, Article 15(1)(h) of the GDPR should be interpreted to mean that the credit information agency must provide the calculation method used to establish the credit score, given that there are no conflicting interests worthy of protection. After acknowledging that protection of trade secrets or intellectual property would in principle constitute a legitimate reason for the credit score agency to refuse to provide the individual with the algorithm used to calculate the score, the Advocate General concluded that such protection can never justify an absolute refusal to provide information because there are appropriate means of communication for understanding and guaranteeing a degree of confidentiality.

Furthermore, the Advocate General stated that the wording in Article 12 of the GDPR, requiring that information be provided in a "concise, transparent, intelligible and easily accessible form, using clear and plain language" excludes the alternative that the agency should be obliged to provide the algorithm as such. The Advocate General proposed that the agency should be obliged to provide the individual data subject with explanations sufficiently detailed so that the individual could challenge the decision. According to the Advocate General, this would mean information about the method used to calculate the credit score and the reasons for the result—notably, information on factors considered and their respective weight at an aggregate level.[59]

According to Article 29 Working Party, it is good practice to provide the data subject with information regarding automated decisions in accordance with what is required in relation to decisions that meet the definitions in Article 22 of the GDPR, even for such automated decisions that do not qualify as being based solely on automated processing or as having legal or similarly significant effects.[60]

In the AIA, the goal for the explainability of AI systems is expressed through the notion that, in relation to high-risk AI systems, the requirements of transparency and human oversight should be met at all times. Furthermore, Article 86 of the AIA includes the right to an explanation for all decisions made with the use of AI. Following the article, all affected people subject to a decision made by the deployer based on a high-risk AI system, having legal effects or in similar ways in a significant way affects the person in a way that, according to them, has a negative impact on their health, safety, or fundamental rights, should have the right to clear and meaningful explanations of the role of the AI system in the decision-making process and the most important parts of the decision taken. In the article, which bears significant similarities with the wording of Article 22 of the GDPR, there are exceptions entailing that such explanations must not be given if the use of AI systems is exempt from the obligation in accordance with union or national law in accordance with union law. Furthermore, it is stated that the article is only applicable to the extent that the right in the article is not provided elsewhere by union law.

Several articles in the PFD include obligations for platform companies to provide information and explanations for people performing work for the platform companies. As an example, Article 9 includes transparency requirements regarding algorithmic management systems in relation to their use and in relation to the categories of data and main parameters taken into account. This article further states that the platform company should inform people performing work as well as their representatives of the grounds for decisions affecting the accounts or contractual status, payment, or "any decision of equivalent or detrimental effect." Article 11 says all Member States shall ensure that people performing platform work have the right to obtain an explanation from the platform for any decision taken or supported by an ADM system. In the article, it is further stated that the explanation be presented in a transparent and intelligible manner, using clear and plain language.

Accountability

All the legal acts discussed in this paper have internal structures governing sanctions. In the GDPR, in relation to the use of ADM Article 22 entails a right for the data subject to obtain human intervention, a right to express their point of view, and a right to contest the automated decision. Recourse to these remedies is limited to cases where the ADM is based on the exception of consent or contractual necessity. In relation to ADM authorized by law in a Member State or EU, the article does not specify what remedies the data subject is entitled to. However, in accordance with the clarifications provided by the CJEU in the SCHUFA Holding (Scoring), the same remedies must be provided in accordance with the law basing the exception from the prohibition. Furthermore, Articles 77, 78, and 79 of GDPR provide data subjects with the right to lodge complaints when the data subject considers that their data has been processed in conflict with the regulation. The GDPR has been criticized for not being suitable for the protection of rights of workers, given its focus on data subjects as individuals.[61] It should be mentioned, however, that Article 80 entails a possibility for an individual data subject to let an organization represent them in a process where the rights of the data subject in accordance with the GDPR have been infringed.[62] In accordance with Article 82, a data subject has the right to compensation for damage suffered as a result of an infringement of the data protection provided in the regulation.

There are rules on sanctions in the AIA, but in general, this regulation lacks remedies for individuals affected by wrongful use of AI systems. This is related to the notion that the regulation is focused mainly on the relation between actors responsible for the use of AI systems and surveillance authorities. However, Article 85 of the AIA includes a right for any person to file a complaint with the authorities responsible for adherence to the rules in the AIA.

In the PFD, several articles include possibilities for affected workers to seek redress. Of particular importance is Article 11, which entails the right to human review of certain decisions made with the use of automated systems. These decisions affect the account or contract of the person performing work, the payment, or "any decision with similar effects or any other decision affecting the essential aspects of the employment or other contractual relationships." If such a decision "infringes the rights of a person performing platform work," the platform company is obliged to rectify the decision, or if that is not possible, the platform company should offer compensation. There are, in addition, other articles regulating possible routes of recourse for affected workers. Article 18 provides the right to redress, and Article 21 entails a possibility for courts to obtain relevant materials and evidence needed to establish the facts in a process based on the PFD.

Reflections

The Prohibition in GDPR Article 22 Is Far-Reaching in the Algorithmic Management Context

What can be concluded from the above regarding the legality of ADM in the employment context is that the prohibition in Article 22 of the GDPR applies. It is evident that the AIA does not impact the applicability of this prohibition, given that it is established in one of the first pages of the more than 400-page-long document that (on the relation between the AIA and the GDPR) "it is also appropriate to clarify that data subjects continue to enjoy all the rights and guarantees awarded to them by such [u]nion law, including all the rights related to solely automated individual decision-making, including profiling."[63] After the SCHUFA Holding (Scoring) ruling, one can on good grounds argue that the definition of a decision based solely on automated processing encompasses a broad array of decisions where a human has used automation as part of the decision-making process.

In acknowledging the severe effects that many managerial decisions have on workers, the requirement that significant effects of the automated decisions exist in order for them to be prohibited will most likely also be met in most ADMs used in algorithmic management. Given the position taken by Article 29 Working Party on the suitability of basing an ADM on explicit consent in the employment context, and the strict interpretation of what constitutes contractual necessity, an employer who wants to avoid the risk of being fined will probably avoid basing ADM on that exception.

This leaves us with the last exception—that a Member State or EU law allows for ADM, provided that such law meets the required criteria of suitable safeguards. Article 88 of the GDPR establishes that deviations from the GDPR in the employment context may be provided only through Member State law or collective agreement if such rules are more specific to ensure the protection of the rights and freedoms in respect of the processing of employees' personal data and provided that such law establishes certain safeguards. Therefore, it is questionable whether the GDPR allows for basing ADM in the employment context on the law exception. This is because it is difficult to see how a law or collective agreement allowing extended use of ADM in the employment context would meet the requirements of Article 88.

In this regard, it should be noted that the rules targeting ADM in algorithmic management in PFD include limitations of what type of data can be used in ADM systems and provide additional information requirements for the use of such systems. Therefore, the PFD does not allow for extended use of ADM in relation to the GDPR. If the conclusions drawn above hold true, the employer who wants to use ADM in an algorithmic management context has two options. The first is to make sure that the human impact in the decision-making procedure is qualitative enough to avoid the decision being considered based solely on automated processing. The second option is to make sure that decisions made with the use of ADM are only those that do not have significant effects on workers.

In sum, in the EU, the only decisions that can legally be made with the use of ADM in the employment context are those that (1) are made with a human in the loop whose impact on the decision is not merely rubber stamping or (2) affect workers in only nonsignificant ways.

The Legal Framework on Algorithmic Management Potentially Constitutes a Novel Form of Limitations on the Managerial Prerogative

The managerial prerogative was described above as a residual competence (taking the three different forms of being *material*, *procedural,* or *personal*) entailing the scope of decision-making where no law, collective agreements, or regulation in individual contract limit the unilateral power of the manager.

The legal framework discussed in this paper, applicable to the use of ADM in managerial functions, seems to entail that a new form of managerial competence (and limitations thereof) is crystalized—that is, the competence to decide how the managerial prerogative should be exercised. In other words, the manager cannot freely decide to use ADM systems to make managerial decisions on matters of significant effect on the employees. This limitation differs from those of the procedural competence mentioned above, as this new form of competence includes the limitation that a human manager makes the managerial decisions, and that the human manager cannot decide to outsource management decisions to technical tools. The basis for the development of this new form of limited competence is the interpretation of what constitutes a decision based solely on the automated processing provided in the SCHUFA Holding (Scoring) case. The new provisions in the PFD on the requirements of a human decision-maker in managerial decisions of certain significance can, after the SCHUFA Holding (Scoring) ruling, be considered to reflect a belt-and-braces policy in this regard. However, given that it is still unclear how the exception related to contractual necessity in Article 22 of the GDPR should be interpreted in the employment context, the detailed protective rules in the PFD are valuable for the protection of workers' rights in algorithmic management.

Another aspect of the legal framework on algorithmic management discussed in this paper in relation to the managerial prerogative is that the rules on explanation of ADM in AIA and the PFD potentially entail new limitations on the prerogative through an extension of the scope of decisions that the manager needs to motivate and review. In sum, the effects of the legal framework on ADM in algorithmic management may have the consequence that workers subject to such forms of management are in a better position to obtain explanations for managerial decisions than workers subject to analog management. The motivation behind these extended limitations of the managerial prerogative is presumably that it must be ascertained that workers are not exposed to risks emanating from automation. Against this background, these limitations are justified, and the legal developments described entail that the dystopia of having an algorithm as a boss is prevented (much thanks to successful and important advocacy work during the last years by legal scholars warning for the risks associated with automation of management functions).

The Legal Framework on Algorithmic Management Provides a Promising Infrastructure for the Demand of Workers' Rights

The central question for the scope of the application of the AIA on the use of ADM in managerial functions will depend on how the concept of *infer* should be interpreted in the wording of the AIA. In the end, it will be up to the

CJEU to assess whether the algorithmic management system is to be considered an AI system under the AIA. If the CJEU applies the same extensive interpretation of what constitutes a system that *infer* as it did for what constitutes a system that *decides*, this will impact the number of automated systems used in managerial functions that would be encompassed by the rules in the AIA. If algorithmic management systems are encompassed by the rules on high-risk AI systems in the AIA, it would improve the working environment, as the automated managerial systems must meet certain standards for safeguards. In relation to the applicability of the PFD rules to algorithmic management, the definitional complexity seems to be related to the demarcation of what managerial decisions have significant effects.

Regarding what function the rules based on Article 88 of the GDPR could have in relation to algorithmic management, the potential content of such rules should clarify under what circumstances ADM is allowed based on consent or contractual necessity or to prohibit the applicability of such exceptions as grounds for the use of ADM in managerial functions. In addition, Article 88 could serve as the legal basis for crafting rules inspired by those in the PFD, in order to include people performing work in sectors other than the platform economy, by further safeguarding against harm from algorithmic management practices.

In sum, the legal framework on algorithmic management discussed in this paper is a promising development in terms of strengthening the position of workers subject to such management, and the rules increase the chances of preventing the materialization of risks to workers' rights. There are still ambiguities in certain matters in this regard, one of them being the question of how often workers' representatives should be consulted on the use of AI systems (both in accordance with data protection impact assessment under Article 35 of the GDPR and obligations under the legal framework of information and consultation). Guidance in this regard could potentially be sought in the elaboration in the recitals of the AIA on the requirements for when the responsibility for high-risk AI systems moves from the provider of an AI system to the deployer.

Given the attention paid to providing safeguards for the use of ADM in managerial decisions, it seems that the legal framework governing algorithmic management provides fertile ground and infrastructure for the effectuation of workers' rights. In order for the rights to be efficiently realized—in other words, for risks to be prevented and accountability to be achieved when they have materialized—workers, and their representatives must roll up their sleeves and claim the rights provided to them in the new landscape of work performed under automated management.

Endnotes

[1] Here defined as the application of a class of statistical methods and algorithms usually subsumed under the generic term of "machine learning" to potentially very large and often heterogenous sets of data ("big data").

[2] European Commission Independent High-level Expert Group on Artificial Intelligence 2019:4.

[3] Regulation of the European Parliament and of the Council laying down harmonised rules on artificial intelligence and amending regulations (EC) No 300/2008, (EU) No 167/2013, (EU) 2018/858, (EU) 2018/1139 and (EU) 2019/2144 and Directives 2014/90/EU, (EU) 2016/797 and (EU) 2020/1828 (Artificial Intelligence Act).

[4] European Parliament legislative resolution of 24 April 2024 on the proposal for a directive of the European Parliament and of the Council on improving working conditions in platform work (COM(2021)0762-C9-0454/2021-2021/0414(COD)).

[5] Regulation 8EU) 2016/679 of the European Union and of the Council of 27 April 2016 on the protection of natural persons with regard to the processing of personal data and on the free movement of such data, and repealing Directive 95/46/EC (General Data Protection).

[6] Dignum 2019:3.

[7] OED Online (2021) Algorithm. OED Online. Oxford: Oxford University Press. Algorithm noun - Definition, pictures, pronunciation and usage notes, Oxford Advanced Learner's Dictionary at https://www.oxfordlearnersdictionaries.com/definition/english/algorithm

[8] Lee et al. 2015 :1603.

[9] Nguyen and Mateescu 2019:1.

[10] Adams-Prassl 2019: 139; Polson and Scott 2018.

[11] Adams-Prassl 2022:236.

[12] Kellogg, Valentine, and Angele 2020:367f.

[13] These grounds are: that the data subject has consented to the processing of the personal data for one or more specific purposes or that the processing is necessary for i) the performance of a contract to which the data subject is party or in order to take steps at the request of the data subject prior to entering into a contract, ii) compliance with a legal obligation to which the controller is subject, iii) in order to protect the vital interests of the data subject or of another natural person, iv) the performance of a task carried out in the public interest or in the exercise of official authority vested in the controller or v) the purposes of the legitimate interest pursued by the controller or by a third party, except where such interests are overridden by the interests or fundamental rights and freedoms of the data subject which require protection of personal data, in particular where the data subject is a child.

[14] Case C-634/21 OQ v. Land Hesse, joined party: SCHUFA Holding AG (Scoring) Judgment of the Court (First Chamber) issued 7 December 2023.

[15] n. 13 para 62.

[16] n. 13 para 61.

[17] n. 13 para 63.

[18] Established by Directive 95/46/EC, replaced by the European Data Protection Board established in Article 68 GDPR.

[19] Barros Vale and Zanfir-Fortuna. 2022:9.

[20] Article 29 Data Protection Working Party 2018:13.

[21] Article 29 Data Protection Working Party 2018:23.

[22] Article 29 Data Protection Working Party 2017:7.

[23] Article 7(3) GDPR.

[24] This follows from Article 22(2) b GDPR.

[25] n. 13, p. 66.

[26] Case C-34/21 Hauptpersonalrat der Lehrerinnen und Lehrer beim Hessischen Kultusmnisterium v Minister des Hessischen Kultusministeriums Judgment of the Court (First Chamber) issued 30 March 2023.

[27] Annex III 4 (b) AIA.

[28] Article 3 (1) AIA.

[29] Recital n. 12 AIA.

[30] Recital n. 10 AIA.

[31] European Council of the European Union. 2024. "EU rules on platform work." https://www.consilium.europa.eu/en/policies/platform-work-eu/

[32] Article 2 (1) PFD.

[33] Recital n. 8 PFD.

[34] The PFD includes rules establishing a presumption of the existence of an employment relationship. The rules in the PFD an algorithmic management do, however, cover all persons performing work for a platform company not depending on their contractual status.

[35] Recital n. 49 PFD.

[36] Article 2 (9) PFD.

[37] Coase 1937:390-392.

[38] Anderson 2017:51f.

[39] Rönnmar 2006:61. In addition to reasons of economic efficiency, a focus on property rights of the employer and the implication that the workplace is part of the managerial property has been used as arguments motivating the managerial prerogative, see Davidov 2016:175.

[40] Racabi 2021:87f and Summers 2000:65.

[41] De Stefano 2018:14.

[42] Davidov 2016:175 and Evju 2003:19. The notion that an employer cannot dismiss workers at will can be considered a limitation on the managerial prerogative. The ILO Convention No 158 offers an international standard of what constitutes unfair dismissal. It follows from Article 4 of the convention that termination of employment shall not take place unless valid reason exists. Valid reason is in the article defined as connected with the capacity of the worker or based on operational requirements of the undertaking, establishment or service. In article 5 and article 6 of the convention, reasons that are not valid for termination of

employment are listed. Concerns regarding the privacy of the worker, as well as prohibition against discrimination also constitutes limitations to the managerial prerogative, see Rönnmar 2004:47f and 65.

[43] Evju 2003:20.

[44] Ibid.

[45] This point was made by Nielsen in relation to how the concept of the managerial prerogative is used in an EU context. Nielsen 1996:21..

[46] Malmberg 1997:247.

[47] Arbetsledningsrätt och arbetsskyldighet: en komparativ studie av kvalitativ flexibilitet i svensk, engelsk och tysk kontext, p. 43.

[48] Rönnmar 2004:44.

[49] Article 3 (4) AIA. Deployer is defined as "a natural or legal person, public authority, agency or other body using an AI system under its authority except where the AI system is used in the course of a personal non-professional activity.

[50] Annex III 4 (b) AIA.

[51] Article 3 (3) AIA. Provider is defined as "a natural or legal person, public authority, agency or other body that develops an AI system or a general purpose AI model or that has an AI system or a general-purpose AI model developed and places it on the market or puts the AI system into service under its own name or trademark, whether for payment or free of charge.

[52] Article 43 AIA includes rules on conformity assessment. See Recital n. 64 AIA regarding the scope of the obligations of third-party conformity assessments.

[53] Bathaee 2018:893–894.

[54] Pehrson 2018:26.

[55] Wachter, Mittelstadt and Floridi 2017:96.

[56] Wachter Mittelstadt and Russel 2018:843.

[57] Brkan 2019:106; Mendoza and Bygrave 2017:16–17.

[58] Case C-634/21 OQ v. Land Hesse, Joined party: SCHUFA Holding AG (Scoring), proposal for decision by Advocate General Pikamäe 16 March 2023.

[59] Ibid paras. 53–58.

[60] Article 29 Data Protection Working Party 2018:25.

[61] Calacci and Stein 2023.

[62] Case C-319/20 Meta Platforms Ireland Limited v Bundesverband der Verbraucherzentralen und Verbraucherverbände – Verbraucherzentrale Bundnesverband e.V., Judgment of the Court (Third Chamber) issued 28 April 2022.

[63] Recital n. 10 AIA.

References

Adams-Prassl, Jeremias. 2019. "What if Your Boss Was an Algorithm? Economic Incentives, Legal Challenges and the Rise of Artificial Intelligence." *Comparative Labor Law and Policy Journal,* Vol. 4, no. 1, pp. 123-146.

Adams-Prassl, Jeremias, 2022. "The Challenges of Management by Algorithm: Exploring Individual and Collective Aspects." In: Gyulavári, Tamás, and Menegatti, Emanuele, eds. *Decent Work in the Digital Age.* London: Hart Publishing, pp 231-244.

Anderson, Elizabeth, 2017. *Private Government How Employers Rule Our Lives (and Why We Don't Talk About It)*: Princeton, Oxford: University Press.

Article 29 Data Protection Working Party. 2017. *Guidelines on Consent under Regulation 2016/679.* Brusssels.

Article 29 Data Protection Working Party. 2018.*Guidelines on Automated Individual Decision-Making and Profiling for the Purposes of Regulation 2016/679.* Brussels.

Barros Vale, Sebastião, and Zanfir-Fortuna Gabriela. 2022. *Automated Decision-Making Under the GDPR: Practical Cases from Courts and Data Protection Authorities.* Washington, DC: The Future of Privacy Forum.

Bathaee, Yavar. 2018. "The Artificial Intelligence Black Box and the Failure of Intent and Causation." *Harvard Journal of Law & Technology,* Vol. 31, no. 2, pp. 890-938.

Brkan, Maja. 2019. "Do Algorithms Rule the World: Algorithmic Decision-Making and Data Protection in the Framework of the GDPR and Beyond." *International Journal of Law and Information Technology,* Vol. 27, no. 2, pp. 91-121.

Calacci, Dan, and Stein, Jake. 2023. 'From Access to Understanding: Collective Data Governance for Workers." *European Labour Law Journal,* Vol. 14, no.2. pp. 253-282.

Coase, Ronald Harry. 1937. "The Nature of the Firm." *Economica,* Vol. 4, no 16, pp. 386-405.

Davidov, Guy. 2016. *Purposive Approach to Labour Law.* New York: Oxford University Press.

De Stefano, Valerio. 2018. "Negotiating the Algorithm: Automation, Artificial Intelligence and Labour Protection." *Employment Policy Department Employment Working Paper No. 246.* Geneva: International Labour Office.

Dignum, Virginia. 2019. *Responsible Artificial Intelligence. How to Develop and Use AI in a Responsible Way.* Cham: Springer.

European Commission Independent High-level Expert Group on Artificial Intelligence. 2019. "Ethic Guidelines for Trustworthy AI." Brussels.

European Council of the European Union. 2024. "EU rules on platform work." https://www.consilium.europa.eu/en/policies/platform-work-eu/

Evju, Stein. 2003. "Arbeidsrett og styringsrett – et perspektiv." *Arbeidsrett og arbeidsliv,* Vol. 1 no. pp. 3-32.

Kellogg, Katherine C., Valentine, Melissa A., Christin Angele. 2020. "Algorithms at Work: The New Contested Terrain of Control." *Academy of Management Annals,* Vol. 14, no.1, pp. 366-410.

Lee, Min Kyung, Kusbit Daniel, Metsky Evan, Dabbish Laura. 2015. "Working with Machines: The Impact of Algorithmic Data-Driven Management on Human Workers." *Proceedings of the 33rd Annual ACM SIGCHI Conference,* New York: Association for Computing Machinery. pp. 1603-1612.

Malmberg, Jonas. 1997. *Anställningsavtalet - Om anställningsförhållandets individuella reglering.* Uppsala: Iustus Förlag AB, 1997.

Mateescu Alexandra, and Nguyen Aiha. 2019. "Explainer: Algorithmic Management in the Workplace." *Data & Society.* pp. 1-15. https://datasociety.net/library/explainer-algorithmic-management-in-the-workplace/

Mendoza Isak, and Bygrave, Lee A., 2017. "The Right Not to Be Subject to Automated Decisions Based on Profiling." In: Tatiani Synodinou, Philippe Jougleux, Christina Markou, Thalia Prastitou, eds. *EU Internet LAW: Regulation and Enforcement.* Springer, pp. 77-98.

Nielsen, Ruth. 1996. *Employer's Prerogatives - in a European and Nordic Perspective*, Copenhagen: Handelshöjskolens Förlag.Pehrsson, Emily. 2018. "The Meaning of the GDPR Article 22." *European Union Law Working Papers, Stanford-Vienna Transatlantic Technology Law Forum*, no. 31 pp. 1-31

Polson, Nick, and Scott, James. 2018. *AIQ: How Artificial Intelligence Works and How We can Harness its Power for a Better World.* London: Bantham Press.

Racabi, Gali. 2021. "Abolish the Employer Prerogative, Unleash Work Law." *Berkeley Journal of Employment & Labor Law*, Vol. 43, no.1, pp. 79-138.

Rönnmar, Mia. 2004. *Arbetsledningsrätt och arbetsskyldighet: en komparativ studie av kvalitativ flexibilitet i svensk, engelsk och tysk kontext*, Lund: Juristförlaget i Lund.

Rönnmar, Mia. 2006. "The Managerial Prerogative and the Employee's Obligation to Work: Comparative Perspectives on Functional Flexibility." *Industrial Law Journal,* Vol. 35, no. 1, pp. 56-74.

Summers, Clyde W. 2000 "Employment at Will in the United States: The Divine Right of Employers." *University of Pennsylvania Journal of Business Law*, Vol. 3, pp. 65-86.

Wachter Sandra, Mittelstadt Brent, and Floridi Luciano. 2017. "Why a Right to Explanation of Automated Decision-Making Does Not Exist in the General Data Protection Regulation." *International Data Privacy Law*, Vol. 7, no. 2, pp. 76-99.

Wachter Sandra, Mittelstadt Brent, and Russel Chris. 2018. "Counterfactual Explanations Without Opening the Black Box: Automated Decisions and the GDPR." *Harvard Journal of Law & Technology*, Vol. 31, no. 2, pp. 841-888.

VI. Examining Industrial Relations and Employment Relations

Seeking Reciprocity in Capitalism: Reconsidering the Historical Development of Industrial Relations in Japan

NOBUYUKI YAMADA
Komazawa University

Introduction

There is no doubt that every society in history has experienced capitalism, though the extent of its development has varied. This suggests that capitalism is both omnipresent and universal. While this may seem self-evident, as history has demonstrated its prevalence, the mechanisms and conditions underlying the universality of capitalism remain unclear. Notably, despite the many problems capitalism has caused—such as poverty, inequality, and environmental damage—people continue to favor it. Therefore, it is necessary to explain why capitalism has maintained its universal appeal. Furthermore, globalization and neoliberalism have expanded market relations since the end of the twentieth century, and such expansion can also lead to even more social problems. It is because capitalism is based on the commodification of labor force and the expansion of market relations can promote excluding people whose labor forces cannot be full-fledgedly commodified. These people can consist of people with disabilities, unskilled youth, elderly people, and so forth. Such exclusion has brought more social inequalities and divisions.

It is true that capitalism is initiated by capital or capitalists, who can make benefits from the capitalist system. However, these capitalists are only a minority in capitalist society, and its majority are almost workers, who can suffer from capitalist system. Given that capitalism should be supported by most of society to be prevalent in it, workers can also be beneficiaries from capitalism. It is meant that workers are also given and bestowed in capitalism while they are exploited and repressed. This is a paradoxical situation which this paper attempts to explicate.

The theoretical hypothesis of this paper is that the universality of the transition to capitalism can be further assured by the persistence of pre-capitalist social relations, which are expected to be untied through the transitional process. This persistence can maintain reciprocal relations in society, which Karl Polanyi (1956) identified as one of three economic integrative forms. He also stressed that market exchange relations were so dominant that the other relations—reciprocity and redistribution—were not as influential. According to Polanyi, such a situation can lead to the destruction of society, with which capitalism must contend.

If capitalism can prevent such destruction from occurring, at least for a period, other relationships beyond market exchange should also be preserved in society. This paper posits that such preservation is possible if pre-capitalist social relations, which include paternalistic and communitarian relations, can persist and maintain reciprocal relations within capitalist society.[1] Reciprocal relations can be defined as economically interdependent relationships. To some degree, these relations can be sustained through the preservation of pre-capitalist relations and their replacement[2] of industrial relations.

In short, capitalism can paradoxically be stabilized because pre-capitalist social relations, which should swiftly be replaced with capitalist relations at the end of the transitional period, are preserved. This paper is an attempt to verify this theoretical hypothesis by examining the historical development of industrial relations in Japan. Japan has experienced a type of transition in which dominant classes, such as merchant capitalists and the landowning class in pre-capitalist society, turned into industrial capitalists.[3] The fact that dominant classes in pre-capitalist society evolved into a capitalist class suggests that pre-capitalist relations may persist in the long term, as such a capitalist class is likely to maintain greater intimacy or closeness to those relations.[4]

This paper tries to clarify how reciprocity in industrial relations has been enabled to be maintained, taking into consideration not only the logic of interests of the capitalist class but also that of the working class. This paper addresses management familism (*keiei-kazokusyugi* in Japanese) as a tool to preserve reciprocity in industrial relations. It was not only an ideology on industrial relations but also an institution[5] to stabilize them in Japan particularly before World War II.

Although management familism was not necessarily referred to as a pivotal institution of industrial relations in Japan after World War II, the author believes that its basic function of fostering reciprocity[6] has persisted through various institutions that replaced industrial relations with pre-capitalist relations.

First, this paper defines management familism and reviews the debate on its conceptual validity in Japanese-style industrial relations. As a result, it confirms that management familism was not necessarily able to perform a reciprocal function due to the lack of adequate institutions before World War II.

Second, this paper attempts to examine the development of industrial relations in Japan from the perspective of reciprocity and to explore the formation of reciprocal relations through the alignment of the interests of employers and workers. Both employers and workers have two types of interests—one capitalist and the other pre-capitalist. Such alignment can occur only when the pre-capitalist interests of employers and workers are mediated by adequate institutions and align with each other.

Nevertheless, pre-capitalist relations are ultimately destroyed. If reciprocal relations in capitalism can only persist alongside pre-capitalist relations, they cannot exist after such destruction. Lastly, this paper, referencing the modification of Japanese-style industrial relations, indicates that reciprocal relations are now disappearing and suggests what kind of system post-capitalism might resemble.

Management Familism and Its Consequences

Conception

Management familism was, in a limited sense, advocated from the end of the Russo-Japanese War through the 1920s. First, it was an ideology that aimed to obscure class conflicts and integrate workers in Japan by mobilizing pre-capitalist relations in the ideological sphere. This ideology was regarded as a variant of pre-capitalist paternalism. In Japan, as in other societies, an extended family (*ie* in Japanese) has been a normal type of family since the pre-modern (pre-capitalist) era.

Ie is a kind of community in which people live their lives and engage in productive activities such as subsistence agriculture and petty commodity production. It consists not only of kinship but also of non-kinship relations.[7] It is governed by a head of *ie* or a patriarch, and the *ie* community is predicated on both communitarian and ruling relationships. In other words, relationships in *ie* can only be reciprocal through the medium of paternalism.[8]

Management familism depended on the ideology that companies were akin to families. In this view, employers and workers were considered members of one family. In short, this means that presidents are seen as parents and employees as children within the company. Employers, as parents, should protect their employees, while workers, as children, should serve their employers in exchange for their favor.

The reason employers sought to advocate management familism and construct its endorsed institutions was to employ skilled workers in the long term. They needed to raise productivity and win competition. Particularly after the Russo-Japanese War, many large companies in heavy industry struggled to retain these workers. This was because they not only had a tendency to regard themselves as journeymen (*watari-shokko* in Japanese) and aspire to become independent craftsmen based on their skills and experiences, but they also felt freer to move to other companies due to the economic boom following the Russo–Japanese War (Hyodo 1971: 215-220).

Second, management familism employed specific tools to establish its legitimacy. By the 1920s, some of these tools had already been developed, particularly within large corporations—such as mutual aid associations (*kyosai-kumiai* in

Japanese), company-managed stores, and communal benefit facilities. However, these institutions, as instruments of paternalist legitimization, were considered insufficient.

Consequently, management familism was unable to execute the expected functions for integrating workers. After World War I, significant class conflicts emerged in Japan. Workers at that time demanded recognition of their identities as human beings (*jinkaku-syonin* in Japanese). This indicates that pre-capitalist paternalism, which is essential to management familism, was rejected by the workers.

On the contrary, personalistic control in the labor process, which is also one of the pre-capitalist relations and a part of paternalism, gave rise to resistance from workers against the replacement of industrial relations with pre-capitalist relations. In short, employers in those days sought to realize their pre-capitalist interests to control and integrate workers, while workers aimed to realize their capitalist interests to earn higher wages and oppose personalistic control.

Debates

Many scholars have examined management familism. However, several have argued that management familism is not an adequate concept for characterizing industrial relations or labor control in Japan, even if it was useful for understanding these relations before World War II. Hiroshi Hazama (1964), however, took one of the most affirmative stances in favor of management familism.

He asserted that management familism was an important factor in Japanese-style industrial relations and that it continued to function after World War II, even though it was somewhat molded.[9] He also stated that the Kanegafuchi spinning industry was one of the most remarkable companies that established management familism, which was endorsed by some institutions, such as lifetime employment. It is important to note that he emphasized that management familism was mediated by some modern techniques of labor control (Hazama 1978). In summary, he recognized that Japanese-style industrial relations did not consist solely of pre-capitalist relations.

Many scholars who regard management familism as pre-capitalist relations also indicate that it is only a part of industrial relations in Japan. They assert that industrial relations are fundamentally capitalist, with class conflicts serving as the primary initiator in these relations.

Tsutomu Hyodo (1971: 270-311) stressed that pre-capitalist management familism was not capable of restraining class conflicts because both employers and workers had capitalist interests. He asserted that alternative tools to mitigate class conflicts, such as factory councils, were sought instead of management familism. He regarded factory councils as instruments for industrial cooperation. In short, he recognized that some aspects of industrial relations were replaced by pre-capitalist relations, while also realizing that industrial relations were fundamentally driven by the capitalist interests of both parties. In this sense, he emphasized the limitations of management familism.

On the other hand, Andrew Gordon (2012) emphasized that industrial relations in Japan, even before World War II, were primarily shaped by the interests of employers, workers, and bureaucrats. He also recognized that pre-capitalist management familism was not very effective in restraining industrial conflicts or in the formation of Japanese-style industrial relations. In this sense, his realization highlighted the ineffectiveness of management familism. This is attributed to his lack of affiliation with the Marxist tradition in social sciences in Japan. Therefore, his historical description was not founded on any theoretical frameworks. Excluding any theoretical presumptions, it relied solely on facts and resources.[10]

In comparison, Yutaka Nishinarita (1988: 150–51), as a member of the Marxist school, recognized that pre-capitalist management familism was, to some degree, effective in the formation of industrial relations. He uniquely indicated that pre-capitalist ideology and its institutional endorsements were incorporated into the capitalist interests of employers. In this sense, his statement addressed the subsumption of pre-capitalist management familism into capitalist class conflicts.

As mentioned above, the assessment of management familism was influenced by Marxist tradition. Marxist social sciences conceptualized capitalist development in Japan and aimed to understand its uniqueness within a general theoretical framework. Marxist scholars regarded Japan as a latecomer in the world-system. They correctly recognized that Japan's relative backwardness contributed to its rapid industrialization, even as it retained a large agricultural population.

Particularly, Marxist scholars in *Koza-ha*[11] have emphasized that Japan's development was rooted in its rural background, where pre-capitalist social relations persisted and influenced the developmental trajectory. Research on industrial relations in Japan has also been shaped by this theory. Kazuo Okoshi (e.g., 1952) was notably influenced by the *Koza-ha* theory and advocated for the concept of a migrant type of labor force (*dekasegi-gata* in Japanese).

According to Okochi, a *dekasegi-gata* labor force is a primary labor force that moves back and forth between cities and rural areas. This labor force maintains pre-capitalist social relations in rural areas and shapes the fundamental characteristics of industrial relations in Japan—such as labor forces, labor control, labor unions, and the labor movement. For example, since *dekasegi-gata* labor forces are essentially migrant workers, they work for a short period in specific companies. Consequently, systematic or bureaucratic labor control cannot be established.

Furthermore, they are not particularly interested in the labor movement because they are migrants. Additionally, their unions tend to be corporate-based, as they migrated from specific regions to specific companies and were unable to build cross-industrial organizations. Consequently, Okochi attempted to deduce all the characteristics of industrial relations in Japan from pre-capitalist relations in rural areas.

Okochi's theory was criticized both theoretically and empirically by the scholars mentioned above.[12] They sought to incorporate capitalist relations and the interests of employers into the analysis of industrial relations. In short, in opposition to Okochi, they tended to deduce most issues of industrial relations from the logic of capital. This tendency appeared to be more pronounced among economists than among sociologists and business researchers. Sociologists were more interested in the specifics of Japanese society, particularly in the fields of villages, family, and industry. Many of them also regarded social relations in these areas as pre-capitalist, feudal, or premodern. They aimed to analyze the functions of these relations and to abolish them as feudal relics (*hoken-isei* in Japanese).[13]

Is there any validity to management familism as a conceptual framework for understanding industrial relations in Japan? We need to reconfirm several propositions regarding how the concept of management familism is validated. These propositions are derived from Yamada (2019; 2020a; 2020b).

- Given that Japan is classified as a type 2 society in the historical sociology of industrial relations, pre-capitalist social relations, of which management familism consists, can persist over the long term.
- Employers have an intimacy with pre-capitalist relations and a tendency to replace industrial relations with them while pursuing their (capitalist) interests. In other words, employers can pursue their capitalist interests by transforming industrial relations through pre-capitalist frameworks.
- As a result, employers possess both capitalist interests and pre-capitalist interests, which are contradictory yet compatible with each other.
- Industrial relations are determined not only by the interests of employers but also by those of workers. Workers also have an intimacy with pre-capitalist relations and a tendency to approve the replacement of industrial relations with them when they seek to (re)form reciprocal relations within their companies. In other words, workers can endorse capitalism by forming reciprocity in their companies through the medium of pre-capitalist relations.

This way of thinking suggests that management familism can emerge only when the interests of both employers and workers are aligned. In other words, for management familism to be formed and function adequately, the pre-capitalist interests of employers for capitalist benefits must be in accordance with the pre-capitalist interests of workers for reciprocity.

In the following, attempts at reciprocity in capitalism are reviewed, with a focus on the case in Japan. This task demonstrates that management familism, even in an extended sense, can function as an instrument to realize reciprocity in capitalism. A condition for such a realization is that the interests of both employers and workers are pre-capitalist and aligned. A leitmotiv is only suggested in this paper.

What Makes Management Familism Possible?

In the above, we can confirm that the alignment of both pre-capitalist interests of employers and workers is essential for the construction of management familism. Additionally, certain institutional instruments are necessary to ensure the endorsement of ideological statements. Such statements must be supported by genuine institutions. If these institutions cannot be fully established, management familism cannot serve as an instrument for reciprocity within companies.

To put it another way, pre-capitalist relations are, in a sense, double-edged swords. Management familism emerges as a kind of paternalism. On the one hand, if paternalism can entail adequate institutions, it can foster reciprocal relations. On the other hand, if paternalism only manifests as pre-capitalist personalistic labor control without any favorable institutions, it can give rise to the capitalist resistance of workers[14]. In short, without institutions, the pre-capitalist interests of employers are likely to conflict with the capitalist interests of workers. In the following sections, we will examine the accord and/or disaccord of interests and the endorsed institutions that facilitate such accord, providing an overview of the historical development of industrial relations in Japan.

From the End of the Russo–Japanese War Through the 1920s

As mentioned above, management familism emerged as an ideology with its endorsed institutions after the Russo–Japanese War because employers sought to retain skilled workers longer in their companies and increase productivity. However, management familism was limited in its effectiveness for such aims during that time.

Employers advocating management familism sought to (re)establish pre-capitalist relations, compelling workers to serve their companies in exchange for certain favors. In essence, management familism was a form of paternalism rooted in the metaphor of the family (i.e., management as parents and workers as children). Under this framework, management, as parental figures, demanded obedience from workers, who were expected to comply in return for favors. This approach was grounded in the concept of master–servant companionship (*shujyu-no-jyogi* in Japanese).

If employers could have provided their workers with sufficient benefits by building appropriate institutions, workers would have accepted personalistic labor control, which included management familism. This is because workers had a familiarity with pre-capitalist relations. Many workers moved from rural areas,[15] where such relations were strongly persistent and prevalent.

Nevertheless, it was natural for workers to have capitalist interests in raising wages and improving working conditions. After World War I, the rise of socialism and the establishment of the International Labor Organization (ILO) also affected the labor movement in Japan. For example, *Yuai-kai* (the Fraternity Union) became somewhat oppositional, despite its original cooperative stance.

Consequently, some institutions for managing familism, such as mutual aid associations (*kyosai-kumiai* in Japanese), company-managed stores, and communal benefit facilities, were not effective enough to convince workers to cooperate with employers. In other words, reciprocal relations were not created during that time. The capitalist interests of employers, while utilizing pre-capitalist relations, were not aligned with those of the workers. Workers also asserted their capitalist interests instead of accepting pre-capitalist relations.[16] Alternatively, they proposed their pre-capitalist interests by intending to become craftsmen based on their experiences as journeymen.

Both parties in industrial relations had not only capitalist but also pre-capitalist interests. However, during that time, these interests did not align with each other. For example, the capitalist interests of employers conflicted with those of workers, leading to class conflicts. Furthermore, due to inadequate institutions, the pre-capitalist interests of employers also clashed with those of workers, despite both having pre-capitalist interests.

The result of these disaccords did not lead to the realization of reciprocal relations in capitalist Japan. How were reciprocal relations guaranteed in those days? If reciprocity had not been fully guaranteed, Japanese society could have faced dissolution. However, Japan was not fully industrialized and preserved a wide rural region. In this rural region, as mentioned above, pre-capitalist relations persisted.[17]

Furthermore, a relatively larger population lived in the rural region than in the urban region. Reciprocal relations were likely to derive from *ie* and its entailed pre-capitalist relations. *Ie* realized paternalism, and the moral economy was thought to be formed by patron-client (*Oyabun-Kobun* in Japanese) relations in *ie*.[18] In this way, given that capitalism in Japan in the semiperipheral zone was somewhat limited, Japanese society was capable of guaranteeing reciprocal relations.

From the 1930s Through the End of World War II

As examined above, until the 1920s, management familism did not function effectively in building benign industrial relations. However, pre-capitalist relations did not disappear from Japanese society. Given their origins, employers were inherently connected to pre-capitalist relations. Furthermore, many people still lived in rural areas, where they were influenced by the *ie* system, its paternalism, and the sharecropping system. These rural areas also supplied workers who had familiarity with pre-capitalist relations.

In addition, after the Meiji Restoration, the state presented the ideological metaphor of family (*ie*) and insisted that the entire nation should be the children of the emperor (*tennō*). This ideological stance affected the formation of social relations in all spheres of Japan. Of course, the persistence of pre-capitalist relations did not necessarily imply that industrial relations could be benign simply by replacing industrial relations with pre-capitalist ones.

From the 1930s through the end of World War II, employers were in trouble with the labor movement, which was based on the capitalist interests of workers. They pursued creating alternative institutions to labor unions and forced industrial relations to operate cooperatively. Hyodo (1971: 367–74) asserted that such an idea of cooperation was different from management familism and that the former was capitalist while the latter was pre-capitalist.

Gordon (2012) stressed that industrial relations in those days were not related to management familism and that they were formed through the cooperation of the interests of parties—employers, workers, and the state. He also made much of the intentions of bureaucrats (the state) in forming industrial relations in the period before and during World War II.

Nevertheless, it seems especially important that employers pursued cooperative industrial relations rather than oppositional ones and that bureaucrats sought to integrate workers to win the total war through using the family metaphor. Taking these situations and their background into account, each party in industrial relations was thought to be affected by pre-capitalist relations.[19]

When employers preferred cooperative industrial relations to oppositional ones, they seemed to be affected by paternalism. When bureaucrats sought to integrate workers into the Industrial Patriotic Society (*sangyo-hokoku-kai* in Japanese), they were also influenced by pre-capitalist communitarianism mediated by nationalism. When workers accepted such nationalist industrial relations, they were sympathetic to them, even if they were repressed by the totalitarian state.

Employers in the heavy industry sector sought to establish factory councils instead of labor unions. Their primary aim was to achieve industrial cooperation. However, although factory councils addressed workers' demands—mainly related to fringe benefits—they did not engage in collective bargaining over wages. In short, these organizations were initiated by employers to create industrial relations that were advantageous to them. As previously noted, the establishment of factory councils, even if influenced by the employers' pre-capitalist interests, was ultimately rooted in capitalist motives. Consequently, these organizations failed to foster reciprocity.

It is true that the establishment of the ILO influenced the legal recognition of labor unions in Japan (e.g., Nishinarita 1988: 239–318). However, even though the approval of collective bargaining was partly suggested by the state, most employers attempted to restrain the functions of labor unions. This intention was also rooted in the capitalist interests of employers, mediated by pre-capitalist authoritarianism. Ultimately, labor unions were not legally recognized in Japan before World War II. Authoritarianism could never foster reciprocity.

What about the activities of the state? The Industrial Patriotic Society was established by the totalitarian state. This organization was also unilaterally beneficial to the state and its collaborating employers. Workers were forcefully mobilized to work for a nationalist goal in exchange for no benefits.[20] I In short, the state pursued its imperialist (capitalist) goals by mobilizing pre-capitalist ideology, resulting in workers being merely victims. Therefore, the *tenno-sei* fascist state also established no reciprocity.

In those days, the way reciprocity was guaranteed was the same as in the former age. Until the 1960s, during the high economic growth period, Japan had a predominantly agricultural population. Its nature resembled that of developing countries, and from a world-systems perspective, it was located in the semiperipheral zone.[21] Given that social relations in rural regions were relatively traditional and pre-capitalist, it is likely that reciprocity was also relatively guaranteed.

From the End of World War II Through the Mid-1990s

What about after World War II? During this period, the industrial relations system in Japan was greatly transformed. Labor unions were legally recognized, and workers gained the basic rights to organize, collectively bargain, and dispute. This labor reform could only be implemented while Japan was occupied by the Allied Forces and the fascist state was dissolved. The so-called postwar reform, including labor reform, aimed at dismantling traditional relations and institutions, such as the sharecropping system and large family businesses (*zaibatsu* in Japanese).

The General Headquarters of the Allied Forces (GHQ) realized that the fascist state depended on these traditional systems and caused the war. The GHQ was also thought to share some Marxist ideas from *koza-ha* regarding the characteristics of Japanese society. Therefore, it attempted to conduct a series of reforms to dissolve pre-capitalist institutions.[22] Could these reforms completely succeed in destroying pre-capitalist relations?

Despite these reforms, pre-capitalist relations persisted. Many sociologists, interested in the modernization of Japanese society, argued that feudal relics should be abolished; however, pre-capitalist relations remained as elements that were expected to disappear. In retrospect, this suggests that they had a significant impact on Japanese society.

For example, a large labor survey showed that workers had dual loyalty to their employers and unions (e.g. Matsushima 1962). In those days, modern industrial relations were thought to be oppositional. Therefore, it is quite strange that workers should also be sympathetic to employers even though inherently oppositional to each other. The reason for dual loyalty was explained in the perspective of feudal relics. In short, on the one hand, workers were sympathetic to their unions, which was expected to realize their capitalist interests. Nevertheless, on the other hand, they were also loyal to their employers because of pre-capitalist paternalism.

Such a consciousness of dual loyalty in those days can be regarded as a remarkable indication that pre-capitalist relations were persistent. However, dual loyalty itself was never indicative of reciprocity, as it merely served to ideologically conceal class conflicts. In fact, one of the results of the postwar labor reformation was the activation of the labor movement, through which workers pursued their capitalist interests. This situation was reflected in the case of Ohmikenshi (Ohmi Silk Thread Company) in 1954. The issue at the center of this dispute was the human rights of female workers. These workers opposed the pre-capitalist personalistic relations established in the company, particularly the compulsory dormitory living arrangements for workers.

Could such an upsurge of the labor movement eliminate pre-capitalist relations?[23] In reality, while metaphors for pre-capitalist relations—such as family (*ie*)—were not explicitly used, they were not abolished but rather restructured and institutionalized again after labor unions lost many industrial disputes as late as the mid-1960s. During the period of high economic growth, employers were believed to maintain a connection to pre-capitalist relations, and workers were similarly affected.

One reason is that pre-capitalist relations were largely drawn from rural areas through the significant migration of peasants. In Japan, a drastic capitalist class transformation took place during the period of high economic growth starting in 1955. By 1960, the working class accounted for only 50% of the economically active population (Ohashi 1971: 84–85). In other words, most of the population had been engaged in non-capitalist activities such as self-employment and agriculture. This indicates that Japanese society remained largely agricultural, even after experiencing rapid capitalist industrialization since the Meiji era. From the perspective of class composition, it can be said that Japan completed its transition to capitalism during the period of high economic growth.

Another reason is that some institutions were contingently established to preserve pre-capitalist relations. Furthermore, these institutions became the basis of reciprocal relations derived from pre-capitalist interactions. They are regarded as pivotal factors in Japanese-style industrial relations—lifetime employment and the seniority system (*nenko-sei*). Corporate-based unions can also function to preserve pre-capitalist relations in capitalist enterprises.[24]

When the labor movement in Japan surged after World War II, workers demanded the capitalist transformation of pre-capitalist relations in industrial relations. For example, they called for the abolition of the status (*mibun*) discrimination between clerks (*syokuin*) and workers (*kouin*).[25] In this regard, their demands were similar to those of the prewar period—the recognition of their personalities (*jinkaku-syonin*).[26] In other words, they sought to abolish pre-capitalist personalistic labor control.

These capitalist interests of workers were first met with the capitalist interests of employers. Until the 1950s, large labor disputes recurred in various industries, driven by employers' pursuit of rationalization, such as labor-saving measures. Any accord of capitalist interests between both parties, if it existed, never led to the creation of reciprocity. However, some institutions established by employers' pursuit of capitalist interests were mediated by pre-capitalist interests and/or intimacy, contributing to the formation of reciprocal relations in capitalist firms.

As indicated, these institutions consist of lifetime employment, a seniority system, and personnel assessment (*jinji-kouka*). Lifetime employment was introduced as part of management familism in the 1920s. However, this attempt was not fully realized due to a lack of institutional support and conflicting interests between both parties. Lifetime employment was fully established during the period of high economic growth.[27]

Such economic growth stabilized employment and enabled this institution to establish itself, particularly in large companies. The abolition of discrimination between clerks and workers also contributed to the formation of this institution by making internal promotion possible. Given that industrial relations are primarily initiated by employers, lifetime employment—that is, long-term employment—was considered convenient for the interests of employers.

Furthermore, lifetime employment, in a double sense, reflects the preferences of employers. One reflects capitalist preference, and the other reflects pre-capitalist preference. Regarding capitalist preference, lifetime employment, combined with a seniority system, aims to enhance productivity and efficiency.[28] It has facilitated long-term skill formation and mobilized the loyalty of workers.

As regards pre-capitalist preferences, institutions such as lifetime employment and seniority systems have led companies to perceive themselves as *ie*. As mentioned above, employers have begun to move away from family metaphors such as *ie*, *oyabun-kobun* (parent–children), and *syu-jyu* (master–servant). However, it has been repeatedly indicated that these companies are similar to *ie* and communities (e.g., Mito 1991a; 1991b). Workers are compelled to be diffusively engaged in their tasks.[29] (Iwata 1977) and to give priority to their tasks rather than their other personal activities.

Personalistic relations (labor control) have been conducted through personnel assessment. Japanese-style personnel assessment is a unique institution in which workers are evaluated by their bosses[30] without the announcement of assessment results. Workers are personally assessed and feel compelled to obey their employers if they hope for promotions.[31] In this way, workers are totally beholden to their employers. This In this way, workers are entirely obedient to their employers. This situation is similar to a patriarchal system, in which patriarchs personally rule their members.

Recent research also stressed that companies in Japan are like communities. For example, Nomura (2018: 75–81) asserted that some major companies in Japan are formalized as communities and that these companies receive new workers as members of the community, not only as employees. He also emphasized that companies as communities require workers to devote themselves to their tasks within the companies. While he did not state that Japanese companies as communities consist of pre-capitalist relations or that they are based on management familism, his assertion is similar to ours. He also regarded Japanese companies as consisting of communities as a basis, with rational organizations as a superstructure.

Japanese-style industrial relations is not only an instrument for mobilizing the engagement of workers based on capitalist and pre-capitalist relations; it can also provide workers with more rewards—long-term stabilized employment, internal promotion, periodic wage increases, retirement allowances, and so forth.[32] These benefits are fundamentally predicated on contingent factors that build these institutions. However, they are also based on the preferences of both parties.

It is important to note that these benefits cannot be explained solely through economic rationality. For example, long-term employment can become a burden for employers during recessions. Since retirement allowances are paid to workers upon their retirement, rather than for their work, the total amount can be quite expensive. Such economic or capitalist irrationality may arise from the pre-capitalist interests of both employers and workers.

Employers as patriarchs or masters can expect to support their workers, and workers as family members or children can also expect to be supported in exchange for hard work.[33] In this way, reciprocity in capitalist companies can be established through pre-capitalist relations. Even if the metaphor of family is not used, the nature of relationships in Japanese companies can be familial, and the principle of management familism can be realized. This type of industrial relations reached its heyday from the 1970s through the 1980s, leading to the establishment of a corporate-ruled society.[34] With reciprocity functioning effectively, capitalism in Japan was able to operate successfully.

As mentioned above, the presence of reciprocity has transitioned from rural areas to capitalist firms through Japanese-style industrial relations. In Japan, capitalist firms can embody industrial relations that have replaced pre-capitalist relations. This shift is rooted in the dual logic of interests held by employers and workers, both of whom are connected to both capitalist and pre-capitalist relations. Their pre-capitalist interests align when reciprocity is possible. Nevertheless, such reciprocity has been diminishing since the mid-1990s due to globalization.

The Japan Federation of Employers' Association (1995) published a report on Japanese-style management. This report stated that Japanese companies must reconsider so-called *shushin-koyo* (lifetime employment) and differentiate workers or employment style in three types—the type of workers with long accumulated ability, the type of flexible employment, and the type of workers with high professional ability.

In a word, this report suggested that the type of flexible employment should be increased in volume while the type of workers with long accumulated ability should be decreased and the type of workers with high professional ability should be outsourced. Of course, workers with long accumulated ability are involved in lifetime employment and flexible employment means irregular employment.

This could mean that management familism and its reciprocity will finally be dissolved.[35] If reciprocity is dissolved, what will happen in the future? At that point, as Polanyi stressed, capitalism will face a crisis of dissolution, and a transition to post-capitalism will be pursued. Post-capitalism should be, if it exists, more reciprocal.

Table 1. Historical Development of Industrial Relations in Japan

Period	Characteristics	The Logic of Capital	The Logic of Labor	Interests	The Presence of Reciprocity
Until the end of Russo–Japanese War	Transition period	The pursuit of capitalist interests (profits) by pre-capitalist relations, primitive labor relations	Resistance based on pre-capitalist relations, stagnant formation of the working class	Disaccord	Preservation by pre-capitalist mode of production
From the end of Russo–Japanese War to the 1920s	Attempts at "management familism" and its effects	Productivity rise as capitalist interests and paternalism as pre-capitalist interests	Materialism (wage increase) as capitalist interests and craftsmanship as pre-capitalist interests	Opposition	Pre-capitalist relations (*ie*) in rural areas
From the 1930s to the end of WWII	Attempts at "factory councils" and "Industrial Patriotic Society"	The pursuit of capitalist interests (cooperative integration of workers) affected by pre-capitalist interests, nationalism	Materialism (wage increase) and meritocracy (the recognition of personality) as capitalist interests, the resistance against pre-capitalist and nationalist repression	Opposition	Pre-capitalist relations (*ie*) in rural areas
From the end of WWII to the mid-1960s	Postwar democratization and attempts at Japanese-style industrial relations	Rationalization as capitalist interests and pre-capitalist ideology as paternalism	Materialism (wage increase) and meritocracy (the recognition of personality) as capitalist interests	Disaccord/ opposition	Pre-capitalist relations (*ie*) in rural areas
From the mid-1960s to the mid-1990s	The rise of "enterprise-ruled society" and its dissolution	Strong accumulation as capitalist interests and paternalism/communitarianism as pre-capitalist interests	Materialism (spring offense) as capitalist interests and communitarian reciprocity as pre-capitalist interests	Accord	Pre-capitalist relations (*ie*) in capitalist companies
From the 1990s	"Japanese-style management" and "black companies"	Restructuring as capitalist interests and personalistic labor control as pre-capitalist interests	Materialism and human rights as capitalist interests	Opposition	No reciprocity

Concluding Remarks

This paper is an attempt to clarify the reason people prefer capitalism even though capitalism, particularly in globalization with neoliberalism, has given rise to serious problems. It is also an attempt to explicate the reason why transition to capitalism can be universal. Our basic theoretical hypothesis is that transition to capitalism can be universal because pre-capitalist relations are persistent, not destroyed. It is a paradox. The persistence of pre-capitalist relations enables reciprocity to emerge within capitalism and helps avoid the dissolution of society.

Based on the historical sociology of industrial relations, we first confirmed that pre-capitalist relations, although the extent to which they can persist varies, were present in any society after the transition period ended. Furthermore, we chose the case of Japan, where pre-capitalist relations are believed to be preserved over the long term, as type 1. In this paper, we aimed to review the historical development of industrial relations in Japan.

As a result, both parties in industrial relations—employers and workers—have dual interests derived from capitalist and pre-capitalist relations, and the alignment of pre-capitalist interests of both parties makes reciprocity plausible. In Japan, such reciprocity has been pursued in the process of forming management familism. Even though industrial relations in the postwar era are not typically characterized by management familism, we assert that the basic mechanism of management familism, rooted in pre-capitalist paternalism, has been inherited. The establishment of a corporate-ruled society can be regarded as its heyday.

However, since the 1990s, globalization and neoliberalism have forced employers to abolish pre-capitalist relations. This means that Japanese-style industrial relations were modified, and its preserved reciprocity was dissolved. If reciprocity is not guaranteed and capitalism is faced with a crisis, various attempts at post-capitalism need to be pursued in Japan.

This paper is just a leitmotiv, in which the mechanism of the formation of reciprocal relations in capitalist firms is clarified. Yamada (2024) attempted further research. In addition, a case of type 2 should swiftly be examined.

Endnotes

[1] In other words, redistribution can be maintained through the establishment of the welfare state. It can be regarded as the reproductive mechanism of capitalism itself.

[2] It is to be noted that such replacement is not overall but partial. With the end of the transitional period, industrial relations in the economic sphere should be capitalist–wage relations.

[3] In our attempt, "the historical sociology of industrial relations," three types of transition to capitalism are categorized (Yamada 1996). Japan is a case of type 1. The others are type 2 and type 3. A case of type 2 is the United States, in which subordinate classes—for example, the producing class—turned into the capitalist class. In type 2, pre-capitalist relations can be vanquished in the shorter term because such capitalist class is likely to hold "smaller intimacy" to those relations. A case of type 3 is many developing societies—for example, Malaysia—in which foreign capital initiated capitalist development in the society. In type 3, pre-capitalist relations can be persistent in the longer term or vanquished in the shorter term. In other words, the persistence of pre-capitalist relations can be indefinite because foreign capital is neutral to development in the society. In short, foreign capital does not hold definite interests for the persistence of pre-capitalist relations because it does not necessarily benefit from pre-capitalist relations in the society.

[4] This is because these classes benefitted from pre-capitalist relations.

[5] An institution is defined as a constant pattern of industrial relations. In Japan, the lifetime employment and seniority system (*nenko-sei* in Japanese) is representative among such institutions.

[6] Needless to say, industrial relations are generally capitalist relations, and they consist of exploitation and dominance in the labor process. It means that industrial relations generally include no reciprocal relations. Therefore, reciprocal relations are protected through the persistence of pre-capitalist relations. Generally speaking, social relations in a family are thought to be relatively reciprocal in capitalist society. However, given that patriarchal relations are also present in a family and that the principle of *ie*, as mentioned below, is prevalent in Japan, reciprocity in a family is also limited in capitalism.

[7] For example, servants are also thought to be members of *ie*.

[8] Ruling is not reciprocal in the strict sense of the word. However, pre-capitalist paternalism can give rise to trade-off relations of protection by a patriarch and dependence by members. Consequently, relationships in *ie* can be formed to be relatively reciprocal. In fact, *ie* often had its sharecroppers, and moral economy was built between patriarchs and their sharecroppers. Therefore, depending on the quantity of products, the rent was liable to change annually. For *ie* particularly in the rural region, see Aruga (1966a, 1966b). In retrospect, except for nonclass societies, *true* reciprocal relations have never been connected in history.

[9] He stated that it was rather valid to regard the post-war management in Japan as management welfarism than management familism (Hazama 1977). This is because most companies did not come to use the metaphor of family to describe their management style.

[10] Gordon's research was primarily predicated on five corporations in the heavy industry.

[11] There have been two schools of Marxism in Japan—*Koza-ha* and *Rono-ha*. *Koza-ha* was a group of Marxists based on the book series (*Koza* in Japanese) named the *History of Japanese Capitalist Development*. *Rono-ha* was based on the journal named

Workers and Peasants (*Rono* in Japanese). These schools have debated the Meiji Restoration, which was to start modernization in Japan. Moritaro Yamada of *Koza-ha* (1934) was one of the most influential scholars on industrial relations research.

[12] For example, with the development proceeding, many workers in heavy industry came to be employed and in cities. In short, the migrant type of labor force was dominant only in the silk reeling and spinning industry, particularly at the early age. Nishinarita (1988: 15–16) also categorized five groups of labor forces from various industries in the era of industrial revolution after the Sino–Japanese War.

[13] Aruga and Hazama did not necessarily seek to abolish these relations. In particular, Aruga thought of these relations as specific or national in Japanese society, and he tried to show how prevalent and universal these relations are in Japan.

[14] For example, workers are supposed to avoid arbitrariness with personalistic control and to demand more rule-based bureaucratic control.

[15] By this, it was not meant that these workers were migrants.

[16] Based on the resources of Shibaura-Seisakusho (Toshiba), Gordon (2012: 123) also asserted that workers demanded substantial paternalism as their rights. This can be regarded as their demand based on capitalist materialism.

[17] This means that Japan was located in the semiperipheral zone in the world system (Wallerstein 1979). On the one hand, Japan joined so-called imperialist countries rapidly after the Meiji Restoration. However, on the other hand, Japan had widespread rural areas, in which landowner ruled sharecroppers. In this way, Japan had both the core zone, which was characteristic of industrialization, and the peripheral zone, which was characterized by low wages. In this sense, such semiperipheral nature was also reflected in debates on the nature of Japanese capitalism between *Koza-ha* and *Rono-ha*. For this point, see Yamada (1998).

[18] Such patron–client relations were not only tied among *ie* members but also between patriarchs and sharecroppers. According to Aruga (1966a, 1966b), many sharecroppers were likely to be originally members of *ie* such as branches (*bunke*) and branches of servant families (*bekke*).

[19] The same situation seems to have been true of the Factory Law. In Japan, the Factory Law was enacted in 1911, and it protected younger workers from overwork and health damage. It is interesting that some arguments against the Factory Law were predicated on paternalism (e.g., Gordon, 2012: 70–71). In short, employers asserted that industrial relations in Japan were based on master–servant compassionship and that workers were originally protected. Therefore, the Factory Law was not requisite in industrial relations. This showed that their opposition was not capitalist but pre-capitalist.

[20] Industrial Patriotic Society was regarded as belonging to a corporatist organization, as in Nazi Germany. Other fascist states like Germany and Italy, which took power in the name of socialism, tried to bestow workers with some benefits. However, as Nishinarita (1988: 427) indicated, the fascist state in Japan was the most exploitative among the fascist states because it gave the least to workers. Therefore, industrial conflicts were not repressed in this period.

[21] In other words, Japan has climbed up to the core zone only during the high economic period. It is very strange that Japan should hold large agricultural population as late as in the 1960s.

[22] For example, the literature influenced by *koza-ha* was published in the United States (e.g., Norman 1940). Such literature was likely to affect the policy of the GHQ on postwar Japan.

[23] One of the reasons why pre-capitalist relations should be abolished, even if it can be beneficial to employers, is class conflicts. However, it is required that employers should prefer abolition to persistence so that pre-capitalist relations are finally extinguished.

[24] By that, it is meant that corporate-based unions gradually come to be integrated by enterprises because they only organized employees of specific enterprises. They tended to function as an instrument for the formation of communitarian relations.

[25] Until the end of World War II, clerks and workers belonged to different statuses, and various discriminations between them were implemented. Such discriminations included not only wages but also facilities such as cafeterias, gates, and the like. In short, clerks and workers were required to use different cafeterias and gates.

[26] Inherently, as late as in the 1920s, workers were generally treated like poor citizens in cities even though their wages rose, particularly in large companies in heavy industry (Hyodo 1971: 316–20; Nishinarita 2007: 113–61).

[27] As is well-known, lifetime employment has been limited. It only includes male regular workers in large companies. In other words, female workers, irregular workers, and workers in small- and medium-sized companies are not included. This means that many workers have not enjoyed reciprocal relations, which lifetime employment can provide. However, employment practices in large companies have been recognized as social norm, and they have also been regarded as a common goal (Nomura 2007: 106–09).

[28] This assessment can be retrospective and just a consequence of long-term practice. Therefore, there can be doubt that employers preferred to institutionalize this type of employment based on their capitalist interests.

[29] It is well-known that they do not have the concept of "job" in Japanese companies. In other words, workers are not allocated to specific jobs, and they are expected to perform any tasks. This can also enable multiskilled workers to be used through rotation.

[30] The personnel department also arranges the results of assessment in order to use them efficiently for promotion and raises. Such arrangements are often arbitrary.

[31] This assessment was combined with meritocracy (*noryoku-shugi*), which is considered to be modern or capitalist. Capitalist interests of employers were tied with their pre-capitalist interests. In other words, when employers sought their capitalist interests, they were always wedded to their pre-capitalist interests. This is because the cultural way of thinking cannot be easily altered.

[32] These rewards were bestowed unilaterally by employers. They were earned by workers based on capitalist interests. For example, annual collective bargaining, called spring offense (*shunto*), brought about wage increases.

[33] Tojo (2005: 154; 2016) also stressed that the basic principle of Japanese society was different from that of western society. The former is feeding, and the latter is contract.

[34] Corporate-ruled society was characteristic of Japanese society from the 1970s through the 1980s, when people were integrated in enterprises and the activities of enterprises were regarded as top priority.

35 This point is typically reflected in "black companies" in Japan. Yamada (2018) considered that "black companies" reproduced pre-capitalist relations.

Acknowledgments

This paper is based on research project "the prospect of post-capitalism and the origin of capitalism: theoretical and historical research for the reconstruction of the theory of transition" funded by the Japan Society for the Promotion of Science (JSPS).

References

Aruga, Kizaemon. 1966a. *The Family System and Sharecropping in Japan, Vol. 1* (in Japanese). Tokyo: Ochanomizu Shobo.

———. 1966b. *The Family System and Sharecropping in Japan, Vol. 2* (in Japanese). Tokyo: Ochanomizu Shobo.

Hazama, Hiroshi. 1964. *Research on the Japanese History of Labor Management: The Birth and Development of Management Familism* (in Japansese). Tokyo: Diamond Press.

———. 1977. *The Lineage of Japanese-style Management* (in Japanese). Tokyo: Bunshindo.

——— .1978. *The Basis of Industrial Peace in Japan* (in Japanese). Tokyo: Waseda University Press.

Hyodo, Tsutomu. 1971. *The Historical Development of Capital-Labor Relations in Japan* (in Japanese). Tokyo: The University of Tokyo Press.

Iwata, Ryushi. 1978. *The Principle of the Formation of Japanese-Style Management* (in Japanese). Tokyo: Bunshindo.

Japan Federation of Employers' Association. 1995. "Japanese-style Management" in *The New Age: Shifting Direction and Its Tangible Methods* (in Japanese).

Gordon, Andrew. 2012. *The Evolution of Labor Relations in Japan*, 1853-2010 (Japanese extended version). Tokyo: Iwanami Shoten, Arranged by Harvard University Press.

Matsushima, Shizuo. 1962. *Japanese Characteristics of Labor Control and their Changes* (in Japanese). Tokyo: Diamond Press

Mito, Tadashi. 1991a. *The Logic of Ie, Vol. 1: Introduction to the Theory of Japanese-Style Management* (in Japanese). Tokyo: Bunshindo.

———. 1991b. *The Logic of Ie, Vol. 2: The Formation of Japanese-Style Management* (in Japanese). Tokyo: Bunshindo.

Nishinarita, Yutaka. 1988. *Research on the Historical Development of Capital-Labor Relations in Modern Japan* (in Japanese). Tokyo: The University of Tokyo Press.

———. 2007. *Labor History in Modern Japan: The Logic and Verification of the Formation of Labor Force* (in Japanese). Tokyo: Yuhikaku.

Nomura, Masami. 2007. *Japanese Employment Practice: An Attempt at a Total Vision* (in Japanese). Kyoto: Minerva Shobo.

———. 2018. *Why Does Karoshi/Karojisatsu Happen in "Excellent Companies? Japanese Companies as "Black and White Companies"* (in Japanese). Kyoto: Minerva Shobo.

Norman, E. Herbert. 1940 [1993]. *Japan's Experience as a Modern State: Political and Economic Problems of the Meiji Period.* Tokyo: Iwanami Shoten.

Ohashi, Ryuken. 1971. *The Class Composition in Japan* (in Japanese). Tokyo: Iwanami Shoten.

Okochi, Kazuo. 1952. *Japanese Labor Movement at the Dawn of Modern Era* (in Japanese). Tokyo: Iwanami Shoten.

Polanyi, Karl. 1955. *The Great Transformation*. NY: Basic Books.

Shoji, Kokichi. 2020. *The Sociology of Social Change in the 21st Century: Towards the Social Recognition of Sovereignty*. Tokyo: Shinyosha.

Tojo, Yukihiko. 2005. *Modernity, Labor, and Civil Society: The Historical Cognition on Modern Japan*, Vol.1 (in Japanese). Kyoto: Minerva Shobo.

Tojo, Yukihiko. (ed.) 2016. *The Rise of "Labor Force" and Contemporary Civil Society: The Historical Cognition on Modern Japan* Vol. 2 (in Japanese). Kyoto: Minerva Shobo.

Wallerstein, Immanuel. 1979. *The Capitalist World-Economy*. NY: Cambridge University Press.

Yamada, Moritaro. 1934. *The Analysis of Japanese Capitalism* (in Japanese). Tokyo: Iwanami Shoten

Yamada, Nobuyuki. 1996. *Historical Sociology of Industrial Relations: An Attempt at the Theory of Multiple Capitalist Development* (in Japanese). Kyoto: Minerva Shobo.

——— 1998. *Class, the State, and the World-System: A Macro Sociology of Industry and Social Change* (in Japanese). Kyoto: Minerva Shobo.

——— 2018. "How Can Three Questions on Modernity Be Answered? The World-System Theory and Japanese Experiences." *International Journal of Japanese Sociology*. No. 27, pp. 55-69.

——— 2019. "What Can Make Capitalism Rise? The "University" of Transition (in Japanese)." *Komazawa Sociological Review*. No. 53, pp. 129-153.

——— 2020a. "What Can Make Capitalism End? Towards a Revised Theory of Transition (in Japanese)." In Shoji, Kokichi (ed.), pp. 83-101.

——— 2020b. "The Social Background of Industrial Relations : The Influences of Rural Social Relations in Japan (in Japanese)." *Komazawa Sociological Review*. No. 55, pp. 51-74.

——— 2024. *Japanese-Style Industrial Relations and "Moral Economy": The Reciprocity in Capitalism*. Kyoto. Minerva Shobo.

VII. Union Power, Satisfaction, and Composition

Labor Intellectuals

DAVID JACOBS
American University

This brief essay is a first step in a developing project on "labor intellectuals"—labor scholars and policymakers who sought to develop and implement a sustainable model for labor power in the United States. I have in mind such individuals as Frances Perkins, FDR's Secretary of Labor, left-wing New Deal critic Mary Van Kleeck, Walter Reuther's strategist Nat Weinberg, labor lawyer, and historian Staughton Lynd, among others. The emphasis is on labor advocates rather than neutrals.

In *New Men of Power,* C. Wright Mills (1949) defined labor intellectuals as researchers associated with unions or labor-based organizations. One fundamental axis upon which to compare these thinkers is their perspectives on the Wagner Act framework. While Perkins built on her experience in New York State regulating labor standards and played a leadership role in New Deal labor policy, confident that Senator Robert Wagner could devise policies to protect workers' rights, Russell Sage Foundation researcher Mary Van Kleeck opposed the Wagner Act and predicted that its support for labor would not endure. Both Perkins and Van Kleeck learned of workers' material conditions in the settlement houses. They had very different assumptions about the capacity of government to intervene on behalf of workers. It is ironic that Van Kleeck and AFL founder Samuel Gompers were superficially in agreement about the elusiveness of pro-labor government activism.

Nat Weinberg, aide to UAW president Walter Reuther, was an important architect of UAW social unionism, devising a far-reaching social democratic agenda. While the UAW ultimately abandoned its campaign for national planning and input on price and product, Weinberg and Reuther continued to embrace collective bargaining for social change and advanced labor in the era of the Great Society. Weinberg died in 1985 after the Reagan era had put an end to labor's forward march.

Staughton Lynd, a US historian, saw a labor movement in decline in the 1980s and resolved to defend the rank and file as a labor lawyer. He moved to Youngstown, Ohio, to fight plant shutdowns in steel. Rather than focus on refinements in labor law, Lynd recommended continuous mobilization. He shared Van Kleeck's skepticism about legislative solutions.

A "critical biography" template informs this project (Jacobs 2007). This approach identifies five emphases for research: the subject's values and background, considerations of personal identity, important roles, contributions to theory and practice, and the social and historic context. Perkins and Van Kleeck emerged from progressive social work and shared a nonconformist identity. Weinberg was a graduate of the Brookwood Labor College. Lynd was a radical historian and lawyer, the son of sociologists Robert and Helen Lynd.

At a moment of renewed labor mobilization, it is appropriate to revisit the stories of historic labor activists who have struggled with governments and enterprises resistant to calls to democratize and humanize.

References

Jacobs, David. 2007. "Critical Biography and Management Education." *Academy of Management Learning and Education* 6 (1): 104–108.

Mills, C. Wright. 1949. *New Men of Power*. New York: Harcourt, Brace.

Employee Voice and the Digitalization of the Workplace: Positive or Negative for the Company and Its Employees?

ITARU NISHIMURA
Chuo University Faculty of Commerce

HIDEKI S. TANAKA
Doshisha University Faculty of Policy Studies

Abstract

This study aims to contribute to further research on the impact of labor–management communication on improving business and employee life enrichment in digitalization. The study focuses on the impact of labor–management communication, employees' voice forms, and their combination (interaction) on firms' productivity and competitiveness, innovation, and employee life enrichment. By analyzing survey data collected from business sites in Japan, the study reveals that labor–management communication enhances the effectiveness of technology adoption, and a good combination of collective and individual voices may enhance the effectiveness of technology adoption. Furthermore, in some aspects, collective labor–management communication, as represented by trade unions, contributes to improving companies' competitiveness and employees' lives following the introduction of digital technologies. The results indicate that some effects depend on the form of the voice. It has been suggested that it is essential to communicate and consider which form should be used. The results also suggest that the same technology can have divergent outcomes depending on the extent and form of employees' voices.

Introduction

There are two contrasting perspectives on the impact of digitalization on employment. The optimistic view suggests that digitalization spurs job creation, while the pessimistic view warns about job loss (Grimshaw 2020; O'Reilly et al. 2018). Will technological change, as symbolized by digitalization, bring about a uniform workplace transformation? Alternatively, will the effects of technological change vary from one workplace to another, depending on the stakeholders involved? Technological determinism supports the former. However, existing studies on employment relations highlight that the impact of digitalization on the workplace is nuanced and influenced by various factors, such as labor–management communication, management attitudes, and government policies (Kornelakis et al. 2022; Rego 2021). These studies highlight the divergent effects of employee voice on the introduction of new technologies.

However, although there is a growing body of research on the response to the introduction of technology and its impact on employment, there is a pressing need for more research on the effect of employee voice on strengthening the competitiveness of enterprises and improving employees' lives through the introduction of technology. The effect of employee voice on company management and employees' lives has been a primary concern in trade union research (Freeman and Medoff 1987). However, the research on this topic is insufficient. Does the impact of digitalization on competitiveness and employees' lives vary according to the characteristics of labor–management communication? Further studies covering the different phases of technology adoption are required. In recent years, it has also been pointed out that representatives of trade union statements do not necessarily represent employees' will (Barry et al. 2014). In addition, the proliferation of e-voice tools in conjunction with technological developments may, in some respects, make it easier for individuals to communicate their intentions to management, such as internal chats (e.g., Teams and Slack). In employment relations research, the impact of labor–management communication needs to be analyzed considering various forms of voice.

Therefore, this study aims to contribute to further research on the impact of labor–management communication on improving business management and employee working styles during digitalization. It focuses on outcome indicators and voice forms that have not been the subject of previous studies. To achieve this objective, we used questionnaire data on the introduction of digital technology and labor–management communication to conduct an empirical analysis of the impact of labor–management communication on the effects of the introduction of digital technology.

The study reveals that labor–management communication enhances the effectiveness of technology adoption, and a good combination of collective and individual voices may enhance the effectiveness of technology adoption. Furthermore, in some aspects, collective labor–management communication, as represented by trade unions, contributes to improving companies' competitiveness and employees' lives following the introduction of digital technologies.

The remainder of this paper is organized as follows: Section 2 reviews previous studies on relevant employment relationships and presents the focus of the analysis and the variables used in the analysis. Section 3 presents the hypotheses, framework, and variables used in the analysis. Section 4 presents the results. The final section presents theoretical and practical implications.

Previous Research

The Impact of Digitalization

A large body of research has already accumulated on the impact of collective voice institutions on the change digitization brings to the workplace. Studies indicate that the platform economy, as represented by Uber, increases work intensity (O'Reilly et al. 2018; Wood et al. 2018). It has also been noted that this leads to increased employee control and monitoring (Gandini 2019). At the same time, it has been found that platform workers' ability to speak out against the reduction of working conditions depends on the presence of trade unions (Borghi et al. 2021).

Although few studies have focused on nonplatform economies, qualitative and empirical analyses have been conducted on the involvement of trade unions and works councils in digitalization, the extent of their voice, and the consequences on employment. They revealed that trade unions' strategies, their relationship with the government, and management attitudes differ in their responses to the introduction of digital technologies and the impact of their introduction on employment (Kornelakis et al. 2022; Lloyd and Payne 2023a; Lloyd and Payne 2023b; Rego 2021; Stroud et al. 2020).

Based on these empirical studies, it can be argued that the introduction of technology does not lead to a single outcome and that the differences are driven by the industrial relations systems of countries rather than by industry characteristics.

The above studies suggest that the impact of technological change is by no means deterministic and that its consequences depend on the attitudes and actions of actors in the industrial relations system (Kornelakis et al. 2022). Previous studies have focused on the increased control and monitoring of workers associated with introducing technology and its impact on employment. Digitalization is also expected to bring about changes in competitiveness and enhance enterprises and the lives of their employees. The impact of labor–management communication on the competitiveness of enterprises and the improvement of employees' lives due to digitalization has not yet been sufficiently accumulated. Although this is an important topic in employment relations research, it is also necessary to provide evidence of the effects of employee voice on the competitiveness of enterprises and the lives of employees, which have not been questioned in digitalization.

Effectiveness of Voice

The impact of employee voice on management and employees has received considerable attention in industrial relations research. The question "Do trade unions contribute to corporate management?" has been a theme in industrial relations research since the 1980s. Freeman and Medoff pointed out that while they positively impact productivity, they negatively influence company profitability, physical capital investment, and intangible assets (Freeman and Medoff 1987). Doucouliagos and Laroche conducted a meta-analysis of the economic effects of trade unions and showed that the contribution of trade unions to corporate management varied across countries. The effects of trade unions also differ depending on the indicators considered. A positive correlation was found with productivity in the USA. In contrast, a negative correlation has been found for company profitability in the UK and Japan (Doucouliagos and Laroch 2003; 2009). On the other hand, Doucouliagos et al. (2017), who also conducted a meta-analysis, found no correlation in the US manufacturing sector, a negative correlation in the UK manufacturing sector, and a positive correlation in developing countries in terms of the impact on productivity. In addition, it has been shown that the impact differs depending on industry.

In addition to the impact of trade unions on corporate management, research has also accumulated on the impact of trade unions on employees, finding that trade unions function as organizations that communicate employees' will and contribute to improving their lives. Previous studies have found a negative correlation between turnover (Freeman and Medoff 1984; Heavey et al. 2013) and a positive correlation with wages (Jarrell and Stanley 1990). However, some studies have found a negative correlation with job satisfaction (Freeman and Medoff 1984), whereas others have not identified a correlation (Doucouliagos et al. 2017). An important implication of these studies is that trade unions are likely to make different contributions to corporate management, depending on the characteristics of industrial relations in the workplace. Therefore, an analysis that considers workplace industrial relations is also necessary when examining the impact of employee voice on changes in the competitiveness of enterprises and the lives of employees due to digitalization. Following previous studies, this study addresses variables related to productivity and intangible assets and their impact on corporate management. This study does not address physical capital investment, which aims to examine the impact of introducing technology.

Concerning the impact on employees, existing studies have accumulated data on the impact on wages and employee satisfaction but not on the impact on working hours and leave, which are considered to impact employees' lives. When examining the impact of digitalization on working styles, it is important to consider such indicators as working hours and leave, which are directly related to employees' lives. Examining these issues is highly important in such countries as Japan, where working hours are longer, and taking leave is more difficult than in developed Western countries. Therefore, this study examines the effects of voice by focusing on indicators that affect employees' lifetimes (so-called life enrichment), such as reduced working hours and increased leave.

Forms of Voice

Do trade unions represent employees' views of the workplace? Studies examining the effectiveness of trade unions implicitly assume that union representatives' statements reflect the views of the workforce. However, there is growing interest in individual and collective statements in employment relations research. Employees' statements have been equated with those of trade union representatives, but the validity of this assumption has been questioned (Barry et al. 2014). In this regard, employee voice research in human resource management (HRM) has turned attention to voice. The literature on employee voice shows that it positively impacts organizational effectiveness and employees' lives (Bell et al. 2011; Nechanska et al. 2020). Employee voice in HRM research is relevant to employment relations. The first important aspect is the diversification of the form of employees' voice: HRM research distinguishes between employee voice through collective organizations, such as trade unions (indirect voice) and the direct statements of individuals (direct voice) (Wilkinson et al. 2014). This distinction is important in employment relations research with respect to the representativeness of the voice of employee organizations.

Second, the means of voice are diverse. Technology development has led to the diversification of the means of voice. In addition to face-to-face communication, interest in ways of communicating opinions to companies through e-voice has emerged in recent years, such as e-mail or internal chat tools. E-voice refers to how employees use electronic communication tools to raise issues, provide solutions, or participate in decision-making. These methods include email, chatrooms, online surveys, and social media (Greer 2002; Bernauer and Kornau 2022). While employment relations research has paid a high degree of attention to digitization, it has not been observed that such environmental changes are changing how employees express their opinions. In other words, the existing research on employment relations does not consider the diversity of voices. This study draws on the knowledge of voice research accumulated in the neighboring field of HRM to conduct an analysis that considers diverse forms of voice.

As described above, in employment relations research, while studies on the voices of trade unions and employee representative organizations at the time of technology introduction have accumulated, there has been insufficient analysis of their impact on the effects after the introduction of technology. While research on digitalization and the voice of employees in terms of its impact on their employment and the strengthening of management control and supervision has been accumulating, there is insufficient accumulation on the effects of the introduction of technology on corporate management and the lives of employees, which has been accumulating mainly in trade union research.

In addition, while analyses incorporating the diversity of voice forms have accumulated mainly in HRM, employment relations research on the impact of digitalization has not conducted analyses that consider the diversity of employee voice forms. Therefore, this study analyzes an outcome that has not been addressed in employment relations research on the impact of digitalization—that is, the impact of employee voice on the competitiveness of the company and the lives of employees—by considering various forms of voice.

Frameworks and Variables

Analytical Framework and Hypotheses

This study conducts the following two analyses: Analysis 1 analyzes the impact of the degree of usual labor–management communication on the company's competitiveness and its employees' life enrichment. The hypotheses tested in Analysis 1 were as follows.

Hypothesis 1: **The degree of regular labor–management communication has a positive impact on firm competitiveness/innovation/employee life enrichment.**

Analysis 2 examines, based on what was examined in Analysis 1, whether different types of labor–management communication affect company competitiveness/innovation/employee life enrichment when consultations are held on introducing new digital technologies and whether the effects differ depending on the combination of these factors. This study examines whether the effects on corporate competitiveness/innovation/employee life enrichment differ depending on the type of labor–management communication and whether they differ in combination. Specifically, the following three types of communication were discussed in the previous section: indirect voice, real direct voice, and e-voice. The hypotheses tested in Analysis 2 are as follows.

Hypothesis 2-1: **Indirect voice/real direct voice/e-voice positively impacts firm competitiveness/innovation/employee life enrichment when introducing new digital technologies is explained and discussed.**

Hypothesis 2-2: **The combination of indirect voice × real direct voice/indirect voice × e-voice/real direct voice × e-voice when explaining and discussing the introduction of new digital technologies has different impacts on company competitiveness, innovation, and employee life enrichment.**

Because the study analyzes companies in Japan only, social and legal systems are considered to have a certain degree of uniformity. The analytical model is shown in Figures 1-1 and 1-2.

Figure 1-1. Analysis 1: Analysis model

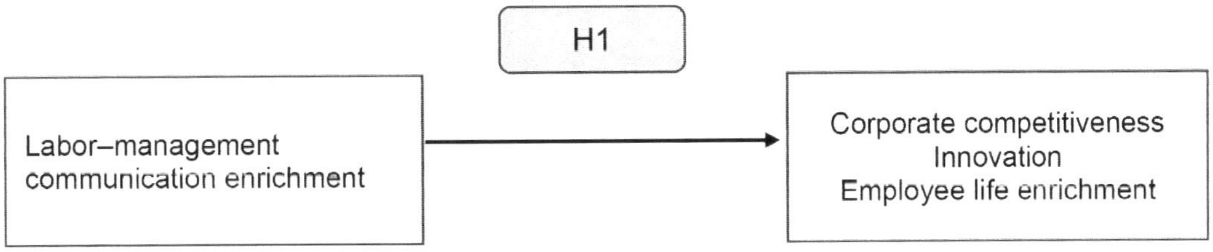

Figure 1-2. Analysis 2: Analysis model

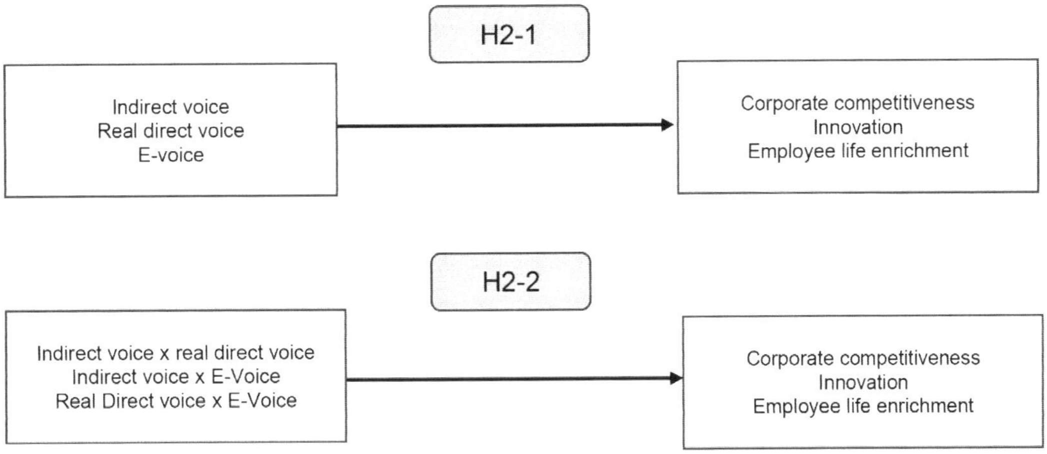

Data

The Survey

The data used in this study are from the "Survey on the Introduction of Digital Technology and Labor–Management Communication" conducted by the Japan Institute for Labor Policy and Training (JILPT). The survey asked the departments in charge of business sites about labor–management communication when companies introduced new digital technologies.

The survey period was from February 27 to May 1, 2023, and covered establishments in Japan with more than 30 employees. The survey was conducted via postal mail. The number of distributed votes was 10,000 (number of undeliverable votes: 108), and the number of collected votes was 1,925, with a collection rate of 19.5%.

The Targets of Analysis

Analysis 1 used a sample with no missing values for the main variables, with a total of 1132 samples. On the other hand, in Analysis 2, only companies that conducted consultations with their employees or provided explanations to them during the introduction of new digital technologies are considered for analysis. Therefore, the sample population for Analysis 2 was

775. After extracting samples with no missing values from the sample population of 775 individuals, the final number of samples for analysis was 685.

Variables

<u>*Explanatory Variables in Analysis 1: Enhancement of Usual Labor–Management Communication*</u>

Concerning enhancing routine labor–management communication, the question "Does your establishment have any of the following means for labor–management communication with employees?" was used. The responses to this question are listed in Table 1. Because this was a multi-answer (MA) question, some respondents chose more than one option. One point was awarded for each selected item, and the total score was used as the usual labor–management communication fulfillment score. The higher the score, the more substantial the permanent labor–management communication, indicating that information is exchanged and shared between labor and management. The minimum value was 0, and the maximum was 8, with a mean of 3.09 and a standard deviation (S.D.) of 1.641.

Table 1. Usual labor–management communication items

<table>
<tr><td>Collective bargaining with trade unions</td><td rowspan="2">Information can be provided via internal newsletters, internal noticeboards, and e-mails</td></tr>
<tr><td>Consultations in labor–management consultative bodies</td></tr>
<tr><td>Formation of a specialized organization to carry out the initiatives</td><td rowspan="2">Communication tools using digital technology, such as internal social networking</td></tr>
<tr><td>Nonpermanent meetings, such as roundtables and information sessions</td></tr>
<tr><td>Consultation services (including by e-mail, etc.) and other complaint bodies</td><td>Employee surveys</td></tr>
</table>

<u>*Explanatory Variables in Analysis 2*</u>

The explanatory variables in Analysis 2 are the forms of voice of employees. The analysis of these voice forms follows the discussion in previous study and divides them into three forms: indirect voice, real direct voice and e-voice.

Labor–management communication using collective voice functions, such as trade unions, is considered indirect voice (hereafter, "indirect"). The selection of "collective bargaining with trade unions," "consultation in a labor–management consultative body," and "formation of a specialized organization to carry out such initiatives" was considered to be the use of the indirect form. One point was awarded for each item selected, and the total score was taken as the indirect score (N = 685, min = 0, max = 3, mean = 0.17, S.D. =.411).

Real direct voice (hereafter, "real direct") refers to a form of speaking out through direct (face-to-face) communication (e.g., exchanging opinions or briefings) with supervisors or executives in the workplace. In the present analysis, the real direct form is considered to have been utilized if the respondents selected "holding briefings for employees," "setting up opportunities for executives to explain policies and exchange opinions," and "explanation in daily work conversations." As for the indirect form, one point was awarded for each selected item, and the total score was used as the real direct score (N = 685, min = 0, max = 3, mean = 1.14, S.D. =.786).

E-voice is a form of voice that uses electronic communication tools to allow employees to raise issues, take action on solutions, and participate in decision-making. These methods include emails, chatrooms, online surveys, social media, etc. (Greer 2002; Bernauer and Kornau 2022). An e-voice form was used if the respondents selected the following options: "receiving consultations at a consultation service (including e-mail, etc.) or other complaint handling bodies," "providing information via internal newsletters, internal noticeboards, e-mail, etc.," and "providing information and collecting opinions via communication tools using digital technology, such as internal social networking sites." As for indirect, one point was given for each item selected, and their total score was used as the e-voice score (N = 685, min = 0, max = 3, mean = 0.69, S.D. = .724).

Explained Variables in Analysis 1 and 2: Outcomes

Analysis 1 and 2 explain the three variables: productivity and competitiveness enhancement, innovation, and employee life enrichment.

Concerning productivity and competitiveness enhancement, in the question "Ask about the effects of introducing new digital technologies," the respondents were considered to have achieved productivity and competitiveness enhancement if they had found effects in "improving the efficiency and productivity of routine tasks," "improving the efficiency and productivity of nonroutine and creative tasks," "enhancing the competitiveness of their products and services," and "improving customer satisfaction." Because the respondents were asked whether an effect was observed, one point was assigned if it was observed. These scores were used as productivity and competitiveness enhancement scores (N = 1,132, min = 0, max = 4, mean = 1.11, S.D. =.822).

In the same questions above, innovation was considered to have been created if an effect was observed in "entering a new and different business field," "changing the current business model," and "changing the current corporate mindset." If the firm was effective for each of these variables, it was assigned a score of 1 point, which was used as the innovation score (N = 1,132, min = 0, max = 3, mean = 0.16, S.D. =.434).

In the same questions above, employee life enrichment was considered to have been achieved if an effect was observed in "ensured employees a relaxed and healthy life," "reducing overtime work," "reducing working on holidays," and "promoting annual leave." One point was awarded for effectiveness for each of these, which was used as the employee life enrichment score (N = 1,132, min = 0, max = 4, mean = 0.69, S.D. = 1.064).

Control Variables

In those analyses, factors other than the management communication variables, such as company attributes and policies, may also affect productivity and competitiveness enhancement/innovation/life enrichment. Therefore, these variables were introduced as control variables in the analysis. The reasons for including policy as a control variable are as follows: Relying on the point made in previous studies that the attitude of the company influences the labor–management communication situation of the establishment and its response to the introduction of technology (Rego 2021), it is expected that the company's policy will influence the way the establishment communicates with labor and its attitude toward its employees. Therefore, in the present analysis, the control variables also include factors related to the company's attitude toward employee statements (decision-making and information-sharing policies).

In addition, as Locke, Kochan, and Piore (1995) pointed out, product strategy, specifically quality- and cost-oriented strategies, significantly impacts the employment relationship; therefore, we decided to include the company's production strategy policy in the controlled variables. This study also included a company's production strategy policy as a control variable.

The following variables were entered as control variables in the analyses: Firm attributes include a manufacturing dummy (manufacturing = 1, nonmanufacturing = 0), year of establishment (1 = up to 1919, 2 = 1920–1939, 3 = 1940–1959, 4 = 1960–1979, 5 = 1980–1999, 6 = 2000 onward), headquarters dummy (headquarters = 1, offices or other = 0), employee size (including part-time employment) (1 = 30 or less, 2 = 31 to 50 employees, 3 = 51 to 100 employees, 4 = 101 to 300 employees, 5 = 301 or more employees), dummy with trade union (with union = 1, without union = 0) were entered.

In addition, the company's management policies and strategies, such as how it conducts business and relates to its employees, were set as controlled variables. Service-oriented, bottom-up, and information-sharing dummies with employees were established. For the service-oriented dummy, a dummy variable was set up in which the value was set to 1 if the choice was "close to A" or "rather close to A" for "A: Emphasis on improving the quality of products and services" and "B: Emphasis on reducing costs" (the other options were "rather close to B" and "close to B"). For the bottom-up type, the dummy variable was set to 1 for "A: Emphasizes bottom-up decision-making" and "B: Emphasizes top-down decision-making." The dummy variable for information sharing with employees was set to 1 for "A: Shares information on business and performance with frontline employees" and "B: Does not share information on business and performance with frontline employees." The dummy variable was set to 1 if "close to A" or "rather close to A" was selected.

Analytical Method

Analysis 1 was a hierarchical multiple regression analysis using the above variables as inputs. In Step 1, only the attributes were entered. In Step 2, in addition to the attributes, the usual labor–management communication scores and the presence of consultations were entered. The results of Step 2 were used to test Hypothesis 1. The results of this analysis are presented in Table 2 (see appendix).

Analysis 2 proceeded according to the following steps: In Step 1, the analysis was conducted using only the entered control variables. Step 2 involves independently submitting the indirect, real direct, and e-voice forms. In Step 3, an analysis was conducted in which the three forms were submitted simultaneously to test the effects of each. The effectiveness of the combination of indirect by direct, indirect by e-voice, and real direct by e-voice was then tested. The results of Analysis 2 are presented in Tables 3-1, 3-2, 3-3 and Tables 4-1, 4-2, and 4-3 (see appendix).

Results

Results of Analysis 1

The results of Analysis 1 (Table 2) reveal the following: First, for the main variables, the usual labor–management communication scores significantly positively affected productivity and competitiveness improvement, innovation, and employee life enrichment ($p < .001$). This indicates that the more extensive the usual labor–management communication, the richer the forms of communication, the more positive the impact on firm performance (productivity and competitiveness improvement and innovation) and employees' life enrichment. On the other hand, consultation had a significant positive effect on productivity competitiveness improvement and employee life enrichment ($p < .01$, $p < .05$), but no significant effect on innovation. These results indicate that consultation and explanation between labor and management regarding introducing new digital technologies has a positive effect on shop floor work efficiency and employee psychology, such as increased productivity and employee life enrichment, but has no effect on innovation.

The effects of the control variables are tested in Step 1. However, the variables influencing each model differed for each explained variable and the significance levels were mixed. Only the service-oriented dummy positively affects innovation in both Steps 1 and 2.

Results of Analysis 2

Analysis 2 tested the forms' effects (Hypothesis 2-1) and the interaction terms for each form (Hypothesis 2-2). In the following section, the results are organized according to each explained variable.

Effects on Productivity and Competitiveness

The effects on productivity and competitiveness enhancement are shown in Tables 3-1 and 4-1. Table 3-1 shows that indirect (β =.096, p <.05:), real direct (β =.155, p < 001:), and e-voice (β =.183, p <.001:) each showed a positive and significant effect on productivity and competitiveness enhancement alone. In Step 3, where these three forms were entered simultaneously, real direct and e-voice showed a significant positive impact (Table 3-1; Step 3: real direct β

=.158, e-voice β =.189, both p <.001). The indirect effect is also positive at the 10% level. These results indicate that real direct voices and e-voices positively affect productivity and competitiveness.

A study on the effects of the combination of indirect × real direct/indirect × e-voice/real direct × e-voice (Table 4-1) revealed the following: First, in the case of the indirect × real direct combination, indirect and real direct had a positive and significant effect on productivity and competitiveness (indirect: β =.083, p <.05, real direct: β =.148, p <.001). On top of that, the indirect by real direct interaction term also showed a significant positive effect on productivity and competitiveness improvement (β =.174, p <.05). In other words, the results suggest that a combination of indirect and real direct effects positively affects productivity and competitiveness. Second, in the case of the indirect by e-voice combination, indirect (β =.090, p <.05) and e-voice (β =.180, p <.001) each had a positive effect on productivity and competitiveness improvement alone. However, in Step 3-2, where the indirect e-voice interaction was introduced, e-voice had a positive and significant effect (β =.161, p <.001), but the interaction had no statistically significant effect. These results indicate that e-voicing significantly affects productivity and competitiveness in an indirect e-voice combination. Finally, in the case of the real direct by e-voice combination, real direct (β =.165, p <.001) and e-voice (β =.192, p <.001) each alone had a significant positive. Although they were shown to have a significant positive impact on productivity and competitiveness gains (Steps 2-3), in the model with the interaction input, the interaction was not statistically significant, and the impact of real direct and e-voice alone on productivity and competitiveness gains was also weaker.

Tables 3-2 and 4-2 show the innovation results. The impact of each form alone, indirect (β =.154, p <.001), real direct (β =.119, p <.01), and e-voice (β =.163, p <.001) each had a positive significant impact on innovation alone. When entered simultaneously (Table 3-2, Step 3), indirect (β =.138, p <.001), real direct (β =.115, p <.01) and e-voice (β =.164, p <.001) showed a significant positive impact. The results indicated that all forms of communication had a statistically significant positive impact on innovation improvement.

Analysis of the impact of the combination of indirect and real direct on innovation (Table 4-2, Step 2-1, Step 3-1) showed that indirect and real direct had a positive and significant effect on innovation directly (indirect: β =.145, real direct: β =.106, p <.001). In addition, the indirect × real direct interaction term also showed a significant positive effect on innovation (β =.202, p <.01). These results suggest that a combination of indirect and direct effects positively affects innovation. In the case of the indirect × e-voice combination, indirect (β =.149, p <.001) and e-voice (β =.158, p <.001) each had a positive effect on innovation alone. In addition, in Step 3-2 (Table 4-2), where the indirect × e-voice interaction was fed in, e-voice maintained a positive and significant effect (β =.125, p <.01), the interaction also had a positive significant effect (β =.116, p <.05). In the case of the real direct by e-voice combination, real direct (β =.127, p <.01) and e-voice (β =.169, p <.001) each had a significant positive effect on innovation alone. The real direct by e-voice interaction positively affected innovation (β =.325, p <.001).

Effects on Employee Life Enrichment

The results of the impact of employee life enrichment are shown in Tables 3-3 and 4-3. First, indirect (β =.149, p <.001), real direct (β =.238, p <.001), and e-voice (β =.135, p <.001) each had a positive and significant impact on employee life enrichment alone (Table 3-3). In addition, the results of simultaneous inputs (Table 3-3 Step 3), indirect (β =.122, p <.01), real direct (β =.235, p <.001), and e-voice (β =.143, p <.001). All the forms positively and significantly affected employee life enrichment.

The results of the analysis of the impact of each combination of forms on employee life enrichment are presented in a comprehensive manner (Table 4-3). In the case of indirect × real direct, indirect (β =.128, p <.01) and real direct (β =.227, p <.001) each alone showed a positive significant impact on employee life enrichment. Analysis with an interaction input also showed that the interaction had no significant effect, and only real direct had a positive and significant effect on employee life enrichment (β =.205, p <.001). In the case of the indirect × e-voice combination, indirect (Step 2-2: β =.144, p <.001, Step 3-2: β =.130, p <.01) and e-voice (Step 2-3: β =.119, p <.05, Step3-3: β

=.120, $p < .01$) each showed a positive and significant impact on employee life enrichment alone. However, this interaction was not statistically significant. For the combination of real direct and e-voice, real direct was shown to have a statistically significant positive impact on employee life enrichment in Steps 2-3 ($\beta = .246$, $p < .001$) and Steps 3-3 ($\beta = .148$, $p < .001$). In e-voice, Steps 2-3 also showed a positive and significant impact ($\beta = .148$, $p < .001$). The interaction also positively impacted employee life enrichment ($\beta = .142$), although at the 10% level.

Analysis Results

The analysis results support the following hypotheses: First, the results indicate that rich labor–management communication with employees and labor–management consultations or explanations from management when introducing new digital technologies might effectively enhance firm competitiveness and employee life enrichment. Therefore, Hypothesis 1 is supported.

Both Hypotheses 2-1 and 2-2 are supported. In other words, voice contributes to improving corporate management and employees' lives through the introduction of technology. The analysis results show that certain combinations of voice further enhance specific effects.

Implication and Conclusion

Theoretical Contributions

The academic implications of this study are as follows: The analysis results of this study show that employees' voices have a desirable outcome for both employees and management after the introduction of digital technology. A meta-analysis has shown that trade union voices have a positive or negative relationship with productivity on a case-by-case basis. Contrary to the debate on whether the effect of union voice is positive or negative, this study suggests that communication with trade unions can positively affect productivity and competitiveness. In addition, union voices are considered harmful to intangible assets; however, the results of this study differ from those of previous studies. Innovation is only one aspect of intangible assets, but labor–management communication may favor one aspect (i.e., innovation).

The results also indicate that some effects depend on the form of the voice. It has been suggested that it is essential to communicate and consider which form should be used. The results also suggest that the same technology can have divergent outcomes, depending on the extent and form of employees' voices. Thus, the results support previous research on employment relations, which rejects the technology-determinist position that technology adoption has a single outcome.

Practical Contributions

This study's contributions are as follows: First, the impact of introducing technology on corporate management and employees differs depending on the combination of specific voice forms. This finding highlights the importance of combining several methods to improve corporate management and employees' lives. Trade unions must remember this as they search for more optimal ways of voice.

The indirect voice also positively affects company competitiveness and employee life enrichment. In other words, a collective voice is meaningful. However, given that real direct and e-voice impact these outcomes, we should not stick to indirect forms through collective industrial relations. We should recognize the need to consider the possibility of using individuals (real direct forms between supervisors and subordinates at the workplace) and e-voices.

Conclusion

The first contribution is that the results support the claims of previous studies on employment relations that have assumed a negative technological determinist position. In Japan, outcomes are also positively affected by the quality of labor–management communication in the workplace rather than by industry characteristics. In other words, it can be said that these results show that the factors that make a difference are not industry characteristics but the characteristics of labor–management communication. In other words, the results of this study indicate that when the labor–management communication of Japanese companies is taken as a sample, the tendency may be the same as the claim of case studies accumulated for European countries (i.e., the characteristics of workplace labor–management communication cause differences by country).

This study also contributes to previous research in which the types of voices have been discarded. Variations were observed in the voices. Focusing on each voice function and its combinations significantly contributes to the present study.

However, this study has some limitations. As the sample is only from Japan, it is impossible to determine whether the results of the present analysis are a Japanese characteristic or whether labor–management communication in Japan (and Europe, if Japan has a European-like labor–management communication reality) has global universality. In this respect, as a future development of this study, a comparison with other countries could lead to overcoming the above limitations.

Appendix

Table 2. Results of the impacts of usual labor–management communication

	Productivity and Competitiveness						Innovation						Life Enrichment					
	Step1-1			Step2-1			Step1-2			Step2-2			Step1-3			Step1-3		
	β	t	p	β	t	p	β	t	p	β	t	p	β	t	p	β	t	p
Const.		6.599	***		4.421	***		2.125	*		.912			4.150	***		2.135	*
Manufacturing (Dummy)	-.038	-1.188		-.034	-1.070		.006	.192		.007	.210		-.042	-1.315		-.039	-1.242	
Year of Establishment (Category)	-.012	-.366		-.008	-.235		-.070	-2.134	*	-.069	-2.112	+	-.042	-1.287		-.039	-1.205	
Head Office (Dummy)	-.001	-.017		-.004	-.143		-.010	-.325		-.014	-.464		-.021	-.676		-.026	-.849	
Number of Employee (Category)	.056	1.781	+	.049	1.577		-.033	-1.051		-.039	-1.242		-.022	-.688		-.030	-.958	
Trade Union (Dummy)	.023	.717		-.040	-1.138		.050	1.529		.001	.028		-.038	-1.153		-.106	-3.043	**
Service Strategy	.056	1.755	+	.042	1.349		.077	2.450	*	.069	2.193	*	.049	1.537		.035	1.131	
Bottom Up Startegy	-.013	-.413		-.025	-.782		.000	-.008		-.008	-.244		.035	1.084		.023	.736	
Information Sharing Starategy	.029	.900		.000	.000		.049	1.550		.030	.989		.042	1.313		.013	.395	
Usual Labour-Management communication scores				.182	5.314	***				.133	3.860	***				.192	5.639	***
Consultation				.089	2.841	**				.035	1.117					.075	2.418	*
Adj.R^2		.003			.039			.011			.026			.005			.042	
ΔR^2				.038		***				.017		***				.039		***
N								1035										

+significant at the 10% level.
*significant at the 5% level.
**significant at the 1% level.
***significant at 0.1% level.

Table 3-1. Results of the impacts on productivity and competitiveness

	Productivity and Competitiveness														
	Step1-1			Step2-1			Step2-2			Step2-3			Step3		
	β	t	p	β	t	p	β	t	p	β	t	p	β	t	p
Const.		5.474	***		5.494	***		4.319	***		4.883	***		3.704	***
Manufacturing (Dummy)	-.045	-1.159		-.050	-1.295		-.048	-1.239		-.031	-.793		-.037	-0.969	
Year of Establishment (Category)	-.032	-.784		-.026	-.642		-.030	-.764		-.015	-.385		-.009	-0.234	
Head Office (Dummy)	.009	.234		.009	.243		.012	.315		.012	.317		.015	.412	
Number of Employee (Cetegory)	.046	1.171		.042	1.080		.044	1.154		.023	.588		.018	0.472	
Trade Union (Dummy)	.041	1.024		.026	.644		.048	1.217		.031	.785		.026	.667	
Service Strategy	.089	2.262	*	.085	2.191	*	.090	2.334	*	.088	2.275	*	.087	2.294	*
Bottom Up Startegy	-.006	-.142		-.008	-.193		-.016	-.407		.005	.130		-.007	-.173	
Information Sharing Starategy	.014	.351		.004	.096		.007	.176		-.001	-.024		-.016	-.424	
Indirect				.096	2.450	*							.075	1.960	+
Real Direct							.155	4.097	***				.158	4.236	***
E-Voice										.183	4.767	***	.189	4.979	***
Adj.R^2	.005			.012			.027			.036			.065		
ΔR^2				.009		*	.024		***	.032		***	.061		***
							(Step1 to Step2)						(Step1 to Step3)		
N							685								

+significant at the 10% level.
*significant at the 5% level.
**significant at the 1% level.
***significant at 0.1% level.

Table 3-2. Results of the impacts on innovation

	Innovation														
	Step1-1			Step2-1			Step2-2			Step2-3			Step3		
	β	t	p	β	t	p	β	t	p	β	t	p	β	t	p
Const.		5.474	***		2.013	*		1.147			1.440			0.633	
Manufacturing (Dummy)	-.045	-1.159		-.019	-.485		-.012	-.319		.003	.070		-.007	-0.175	
Year of Establishment (Category)	-.032	-0.784		-.076	-1.898	+	-.084	-2.104	*	-.070	-1.768	+	-.061	-1.559	
Head Office (Dummy)	.009	.234		.009	.235		.011	.280		.011	.292		.014	.369	
Number of Employee (Cetegory)	.046	1.171		-.032	-.821		-.027	-.689		-.046	-1.191		-.052	-1.375	
Trade Union (Dummy)	.041	1.024		.050	1.237		.079	1.993	*	.065	1.638		.048	1.228	
Service Strategy	.089	2.262	*	.059	1.524		.065	1.679	+	.063	1.633		.060	1.579	
Bottom Up Startegy	-.006	-.142		-.016	-.407		-.020	-.525		-.003	-.082		-.014	-.354	
Information Sharing Starategy	.014	.351		.019	.491		.030	.769		.022	.571		.003	.066	
Indirect				.154	3.984	***							.138	3.614	***
Real Direct							.119	3.128	**				.115	3.082	**
E-Voice										.163	4.231	***	.164	4.324	***
Adj.R^2	.011			.032			.023			.035			.066		
ΔR^2				.022		***	.014		***	.025		***	.059		***
							(Step1 to Step2)						(Step1 to Step3)		
N							685								

+significant at the 10% level.
*significant at the 5% level.
**significant at the 1% level.
***significant at 0.1% level.

Table 3-3. Results of the impacts on employee life enrichment

	Employee Life Enrichment														
	Step1-1			Step2-1			Step2-2			Step2-3			Step3		
	β	t	p	β	t	p	β	t	p	β	t	p	β	t	p
Const.		3.858	***		3.896	***		2.221	*		3.396	**		1.771	+
Manufacturing (Dummy)	-.054	-1.389		-.062	-1.609		-.058	-1.534		-.043	-1.118		-.053	-1.425	
Year of Establishment (Category)	-.063	-1.564		-.054	-1.354		-.061	-1.565		-.051	-1.273		-.041	-1.069	
Head Office (Dummy)	.005	.130		.006	.144		.010	.256		.007	.189		.012	.334	
Number of Employee (Cetegory)	-.035	-.898		-.041	-1.053		-.037	-.971		-.052	-1.331		-.059	-1.580	
Trade Union (Dummy)	-.010	-.249		-.033	-.829		.001	.028		-.017	-.437		-.026	-.671	
Service Strategy	.023	.584		.018	0.468		.026	.672		.022	.572		.021	.561	
Bottom Up Startegy	.042	1.070		.039	1.002	*	.026	.688		.050	1.278		.032	.859	
Information Sharing Starategy	.045	1.137		.029	.746	*	.034	.890		.034	.865		.010	.263	
Indirect				.149	3.816	***				.135	3.484		.122	3.241	**
Real Direct							.238	6.394	***				.235	6.379	***
E-Voice										.135	3.484	***	.143	3.823	***
Adj.R^2		.001			.034			.069			.030			.089	
ΔR^2				.023		***	.056		***	.029		***	.089		***
								(Step1 to Step2)						(Step1 to Step3)	
N								685							

+significant at the 10% level.
*significant at the 5% level.
**significant at the 1% level.
***significant at 0.1% level.

Table 4-1. Results of the effects of interaction on productivity and competitiveness

	Productivity and Competitiveness																				
	Step1-1			Step2-1			Step3-1			Step2-2			Step3-2			Step2-3			Step3-3		
	β	t	p	β	t	p	β	t	p	β	t	p	β	t	p	β	t	p	β	t	p
Const.		5.474	***		4.376	***		4.625	***		4.909	***		4.994	***		3.643	***		3.877	***
Manufacturing (Dummy)	-.045	-1.159		-.052	-1.354		-.048	-1.238		-.036	-.927		-.040	-1.037		-.033	-.859		-.035	-.931	
Year of Establishment (Category)	-.032	-.784		-.026	-.642		-.025	-.621		-.010	-.257		-.010	-.258		-.013	-.341		-.013	-.326	
Head Office (Dummy)	.009	.234		.012	.320		.017	.452		.012	.324		.014	.374		.015	.409		.018	.469	
Number of Employee (Cetegory)	.046	1.171		.041	1.076		.036	.942		.020	.511		.019	.482		.020	.535		.021	.557	
Trade Union (Dummy)	.041	1.024		.035	.874		.038	.950		.017	.430		.018	.446		.038	.979		.036	.917	
Service Strategy	.089	2.262	*	.088	2.269	*	.090	2.343	*	.085	2.208	*	.085	2.209	*	.089	2.355	*	.090	2.371	*
Bottom Up Startegy	-.006	-.142		-.017	-.439		-.015	-.392		.003	.079		.003	.071		-.005	-.139		-.009	-.235	
Information Sharing Starategy	.014	.351		-.001	-.037		.001	.037		-.010	-.259		-.009	-.221		-.009	-.235		-.009	-.244	
Indirect				.083	2.121	*	-.056	-.825		.090	2.323	*	.044	.816							
Real Direct				.148	3.904	***	.107	2.576	*							.165	4.422	***	.108	1.906	+
E-Voice										.100	4.698	***	.161	3.875	***	.192	5.050	***	.116	1.701	+
Indirect × Real Direct							.174	2.465	*												
Indirect × E-Voice													.068	1.208							
Real Indirect × E-Voice																			.104	1.323	
Adj.R^2		.005			.032			.040			.042			.043			.061			.063	
ΔR^2				.030		***	.009		***				.002						.002		
N											685										

+significant at the 10% level.
*significant at the 5% level.
**significant at the 1% level.
***significant at 0.1% level.

Table 4-2. Results of the effect of interaction on innovation

	Innovation																				
	Step1-1			Step2-1			Step3-1			Step2-2			Step3-2			Step2-3			Step3-3		
	β	t	p	β	t	p	β	t	p	β	t	p	β	t	p	β	t	p	β	t	p
Const.		1.992	*		1.241			1.540			1.472			1.643			.528			1.828	+
Manufacturing (Dummy)	-.010	-.265		-.020	-.521		-.015	-.383		-.006	-.152		-.013	-.348		.001	.028		-.008	-.202	
Year of Establishment (Category)	-.085	-2.111	*	-.075	-1.903	+	-.074	-1.887	+	-.062	-1.569		-.062	-1.574		-.069	-1.743	+	-.067	-1.717	+
Head Office (Dummy)	.008	.219		.011	.290		.017	.444		.011	.306		.015	.393		.014	.362		.021	.553	
Number of Employee (Cetegory)	-.026	-.662		-.032	-.837		-.038	-1.000		-.051	-1.334		-.053	-1.385		-.048	-1.247		-.045	-1.191	
Trade Union (Dummy)	.074	1.844	+	.056	1.403		.059	1.495		.042	1.053		.043	1.084		.070	1.791	+	.063	1.616	
Service Strategy	.064	1.634		.060	1.571		.063	1.656	+	.058	1.527		.058	1.530		.064	1.683	+	.066	1.751	+
Bottom Up Startegy	-.013	-.322		-.023	-.585		-.020	-.531		-.007	-.170		-.007	-.183		-.011	-.290		-.023	-.592	
Information Sharing Starategy	.035	.900		.015	.397		.019	.486		.007	.183		.010	.247		.016	.414		.015	.390	
Indirect				.145	3.740	***	-.017	-.243		.149	3.884	***	.071	1.330							
Real Direct				.106	2.815	**	.058	1.413								.127	3.392	**	-.049	-.872	
E-Voice										.158	4.137	***	.125	3.031	**	.169	4.431	***	-.065	-.960	
Indirect × Real Direct							.202	2.880	**												
Indirect × E-Voice													.116	2.079	*						
Real Indirect × E-Voice																			.325	4.142	***
Adj.R^2		.011			.042			.052			.055			.059			.050			.072	
ΔR^2				.034		***	.011		**	.046		***	.006		**	.041		**	.023		**
N											685										

+significant at the 10% level.
*significant at the 5% level.
**significant at the 1% level.
***significant at 0.1% level.

Table 4-3. Results of the impacts of interaction on employee life enrichment

	Employee Life Enrichment																				
	Step1-1			Step2-1			Step3-1			Step2-2			Step3-2			Step2-3			Step3-3		
	β	t	p	β	t	p	β	t	p	β	t	p	β	t	p	β	t	p	β	t	p
Const.		3.858	***		2.312	*		2.443	*		3.445	**		3.486	**		1.671	*		2.167	*
Manufacturing (Dummy)	-.054	-1.389		-.065	-1.723	+	-.062	-1.656	+	-.052	-1.339		-.054	-1.395		-.047	-1.235		-.050	-1.337	
Year of Establishment (Category)	-.063	-1.564		-.054	-1.380		-.053	-1.368		-.043	-1.077		-.043	-1.077		-.048	-1.238		-.047	-1.220	
Head Office (Dummy)	.005	.130		.010	.264		.013	.336		.008	.201		.009	.228		.012	.328		.015	.410	
Number of Employee (Cetegory)	-.035	-.898		-.042	-1.105		-.044	-1.180		-.057	-1.469		-.057	-1.483		-.055	-1.464		-.054	-1.436	
Trade Union (Dummy)	-.010	-.249		-.020	-.497		-.018	-.457		-.040	-.995		-.039	-.986		-.007	-.173		-.010	-.258	
Service Strategy	.023	.584		.021	.567		.023	.603		.018	.459		.018	.459		.025	.663		.026	.685	
Bottom Up Startegy	.042	1.070		.024	.643		.026	.671		.047	1.206		.047	1.201		.034	.908		.029	.775	
Information Sharing Starategy	.045	1.137		.021	.556		.023	.597		.019	.493		.020	.513		.022	.576		.021	.564	
Indirect				.128	3.361	**	.053	.785		.144	3.725	***	.119	2.202	*						
Real Direct				.227	6.120	***	.205	5.040	***							.246	6.653	***	.169	3.000	**
E-Voice										.130	3.385	**	.120	2.871	**	.148	3.927	***	.045	.665	
Indirect × Real Direct							.094	1.359													
Indirect × E-Voice													.037	.657							
Real Indirect × E-Voice																			.142	1.822	+
Adj.R^2		.001			.071			.072			.036			.036			.076			.080	
ΔR^2				.072		***	.003			.037		***	.001						.004		***
N											685										

+significant at the 10% level.
*significant at the 5% level.
**significant at the 1% level.
***significant at 0.1% level.

References

Barry, Michael, Adrian Wilkinson, Paul J. Gollan, and Senia Kalfa. 2014. "Where Are the Voices? New Directions in Voice and Engagement across the Globe." In *The Oxford Handbook of Employment Relations*, edited by Adrian Wilkinson, Geoffrey Wood, and Richard Deeg, 522–40. Oxford University Press.

Bell, Myrtle P., Mustafa F. Özbilgin, T. Alexandra Beauregard, and Olca Sürgevil. 2011. "Voice, Silence, and Diversity in 21st Century Organizations: Strategies for Inclusion of Gay, Lesbian, Bisexual, and Transgender Employees." *Human Resource Management* 50 (1): 131–46.

Bernauer, Vanessa Sandra, and Angela Kornau. 2022. "E-voice in the Digitalised Workplace. Insights from an Alternative Organisation." *Human Resource Management Journal* 34 (2): 369–85.

Borghi, Paolo, Annalisa Murgia, Mathilde Mondon-Navazo, and Petr Mezihorak. 2021. "Mind the Gap between Discourses and Practices: Platform Workers' Representation in France and Italy." *European Journal of Industrial Relations* 27 (4): 425–43.

Doucouliagos, Hristos, Richard B. Freeman, and Patrice Laroche. 2020. *Economics of Trade Unions: A Study of a Research Field and Its Findings*. Routledge.

Doucouliagos, Christos, and Patrice Laroche. 2003. "What Do Unions Do to Productivity? A Meta-Analysis." *Industrial Relations: A Journal of Economy and Society* 42 (4): 650–91.

Freeman, Richard B., and James L. Medoff. 1984. *What Do Unions Do?* Basic Books.

Gandini, Alessandro. 2018. "Labour Process Theory and the Gig Economy." *Human Relations* 72 (6): 1039–56.

Greer, Charles R. 2002. "E-Voice: How Information Technology Is Shaping Life within Unions." *Journal of Labor Research* 23 (2): 215–35.

Grimshaw, Damian. 2020. "International Organisations and the Future of Work: How New Technologies and Inequality Shaped the Narratives in 2019." *Journal of Industrial Relations* 62 (3): 477–507.

Heavey, Angela L., Jacob A. Holwerda, and John P. Hausknecht. 2013. "Causes and Consequences of Collective Turnover: A Meta-Analytic Review." *Journal of Applied Psychology* 98 (3): 412–53.

Jarrell, Stephen B., and T. D. Stanley. 1990. "A Meta-Analysis of the Union-Nonunion Wage Gap." *ILR Review* 44 (1): 54–67.

Kornelakis, Andreas, Vassil Kirov, and Patrick Thill. 2022. "The Digitalisation of Service Work: A Comparative Study of Restructuring of the Banking Sector in the United Kingdom and Luxembourg." *European Journal of Industrial Relations* 28 (3): 253–72.

Lloyd, Caroline, and Jonathan Payne. 2023a. "Trade Unions, Digitalisation and Country Effects: A Comparative Study of Banking in Norway and the UK." *European Journal of Industrial Relations* 29 (4): 325–45.

Lloyd, Caroline, and Jonathan Payne. 2023b. "Food for Thought: Robots, Jobs and Skills in Food and Drink Processing in Norway and the UK." *New Technology, Work and Employment* 38 (2): 272–90.

Locke, Richard M., Thomas A. Kochan, and Michael J. Piore. 1995. *Employment Relations in a Changing World Economy*. MIT Press.

Nechanska, Eva, Emma Hughes, and Tony Dundon. 2020. "Towards an Integration of Employee Voice and Silence." *Human Resource Management Review* 30 (1): 100674.

O'Reilly, Jacqueline, Florian Ranft, and Max Neufeind. 2018. "Introduction: Identifying the challenges for work in the digital age." In *Work in the Digital Age: Challenges for the Fourth Industrial Revolution*, edited by Max Neufeind, Jacqueline O'Reilly, and Florian Raft, 1–23. Rowman & Littlefield International

Rego, Kerstin. 2021. "Works Councils and the Digitalisation of Manufacturing: Opportunity or Threat for Their Power Position?" *Economic and Industrial Democracy* 43 (4): 1911–33.

Stroud, Dean, Claire Evans, and Martin Weinel. 2020. "Innovating for Energy Efficiency: Digital Gamification in the European Steel Industry." *European Journal of Industrial Relations* 26 (4): 419–37.

Wilkinson, Adrian, Tony Dundon, Jimmy Donaghey, and Richard B. Freeman (2014). "Employee Voice: Charting New Terrain." In *Handbook of Research on Employee Voice*, edited by Adrian Wilkinson, Jimmy Donaghey, Tony Dundon, and Richard B. Freeman, 3–16. Edward Elgar.

Wood, Alex J., Mark Graham, Vili Lehdonvirta, and Isis Hjorth. 2018. "Good Gig, Bad Gig: Autonomy and Algorithmic Control in the Global Gig Economy." *Work, Employment and Society* 33 (1): 56–75.

IX. Caring for the Caregivers

Trade Union Organizing of Transnational Care Workers in Germany and the United States

GÜLTEN GIZEM FESLI
University of California, San Diego
University of Bayreuth in Germany
Hans Böckler Foundation

Abstract

This working paper briefly examines and summarizes the unionization strategies as well as power resources of transnational care workers in Germany and the United States, focusing on intersectional approaches. The paper addresses the rising demand for care work driven by aging populations, increased female workforce participation, and the outsourcing of care responsibilities. It highlights the regulatory gaps in live-in care arrangements and the potential strategies that German unions can learn from successful intersectional organizing in the United States, particularly through worker centers such as the Domestic Workers Union (DWU). The working paper is structured around identifying existing collective representation strategies in both countries, with a focus on Germany, assessing the incorporation of intersectional approaches by trade unions as well as the lack thereof. Grounded in the fields of comparative industrial relations and the sociology of gender and work, I use intersectionality in care work as an analytical framework as it allows the comprehension of the multifaceted challenges posed by care, gender, labor market, and migration regimes as well as their implications for migrant care workers operating within 24-hour live-in care arrangements. Furthermore, the paper analyzes power resources of care workers that can help labor unions identify the potential power that care workers hold when organizing them despite their precarious employment situation. Preliminary findings indicate that while the United States has a more developed framework for organizing transnational care workers, there are significant challenges and opportunities for German unions to enhance their strategies to unionize transnational care workers.

1. Introduction

The discourse on care work in industrialized countries such as Germany and the United States has gained prominence due to several factors, including aging demographics, the increased participation of women in full-time employment, and the outsourcing of care work in dual-earner households. These trends have led to a shortage of skilled workers in the care sector and a rise in care migration, particularly in home-based care arrangements. Despite the United States role of care workers, their representation by German trade unions has been insufficient. In contrast, the United States has seen notable successes in intersectional organizing since the 1970s and 1980s, particularly through worker centers like the Domestic Workers Union (DWU).

Organizing approaches were and are still crucial as the working and living conditions of transnational care workers are characterized by precariousness and irregularity. According to Dörre (2014), precarious working conditions refer to a non-sustaining employment relationship that lacks societal recognition and hinders integration into social networks, participation, and long-term life planning. On the other hand, irregularity refers to illegal employment (Habel and Tschenker 2022). In the case of transnational care workers in home-based care, precarity and irregularity are evident in social spatial isolation, constant availability around the clock, lack of access to social benefits, the informal nature of employment due to an ethos of care, the merging of living and working spaces in the so-called live-in care arrangement, and problems with residence permits or illegal entry (Karakayali 2010; Kniejska 2016; Schilliger 2013). Efforts to improve the situation of these transnational care workers in Germany and their union organizing are limited due to barriers related to employment, social conditions, and migration policy.

In contrast, union organizing of transnational care workers in the United States, particularly through the DWU, has made progress. The DWU worked with care workers and civil society partners to strategically organize transnational care workers in private households, including those who illegally entered from Mexico and other South American countries into the United States (Benz 2014). Intersectional approaches played an important role in the organizing processes (Benz 2014; Gordon 2005). Intersectionality in care work allows for the analysis of complex forms of social inequality, such as race, gender, and migration in labor conflicts of transnational care workers (Lutz 2011; Theobald 2024). Thus, the aim of this paper is to highlight the recent union organizing of transnational care workers in Germany using intersectionality as a conceptual framework. A special focus is on the successes of union organizing in the United States as German labor unions such as the United Services Trade Union (German: ver.di) are looking to the United States to potentially learn from successful organizing strategies there. A cross-country reflection on the respective social and political-institutional frameworks is of great importance. To date, there is limited research on the union organizing of transnational care workers in German labor sociology (Bomert 2020; Rudolph and Schmidt 2019; Schwenken 2005). Similarly, the question of the significance of organizing new groups of employees for unions and thus potentially strengthening their social and political power through higher membership numbers (Dörre and Schmalz 2014) has hardly been investigated. The research attempts to fill this gap by examining the organizing of transnational care workers in Germany using the power resources approach (Dörre and Schmalz 2014) and analyzing which power resources unions and transnational care workers possess despite their precarious working and living conditions. The paper has the unique focus of examining the union organizing of transnational care workers in German private households, regardless of whether they have legal residence and work permits. The article also focuses on the United States to discuss the transferability of successful and intersectional organizing strategies from the United States to Germany and thus enable more intersectional union work in Germany. The organizing of care workers by the DWU also achieved the organizing of transnational (and Black) care workers from Mexico and South America, who may not have had legal residence permits (Benz 2014). Thus, the paper connects various social spaces and addresses central questions and themes related to labor and intersectional social inequalities.

2. Research Questions

Although there are studies on the (self-)organizing of transnational care workers by Benz and Bomert, they are not sufficient to answer the question of organizing care workers in Germany. Bomert focuses on social work, while Benz exclusively investigates the transformative character of worker centers in the United States. Thus, the paper examines the specific research question of the union organizing of transnational care workers in Germany and the United States. This includes elaborating on how the union organizing of transnational care workers in Germany can succeed despite social, political, and institutional challenges. In order to answer these questions, existing approaches to collective representation of interests and strategies for organizing care workers in the United States and in Germany need to be identified. These questions will be discussed in the following sections. Before doing so, the influence that the welfare state model in Germany and the United States has on labor union organizing needs to be analyzed in the next section as both industrialized countries are regarded as welfare states in the field of industrial relations. Furthermore, the specific welfare state model has an influence on migration regimes that are part of the intersection of race, class, gender as well as migration regarding transnational care workers. The influence of welfare state regimes will also be discussed.

3. The Influence of the German and US Welfare State on Labor Union Organizing

In order to analyze the strategies and associated structural conditions, as well as to assess the transferability of successful organizing strategies, it is necessary to briefly classify the United States and Germany as welfare states (Esping-Andersen 1998) within the industrial relations. According to Artus (2023), Germany is often categorized as a type of social partnership (significant role of collective bargaining parties, conservative welfare state, few comprehensive unions) and sometimes as a type of corporatism (well-organized bargaining parties, inclusion of bargaining parties in state decisions, existence of social partnership advisory bodies). The United States is categorized as a type of liberal-pluralistic market orientation with the characteristics of weak collective bargaining associations, low collective bargaining coverage, and poorly organized as well as decentralized wage-setting systems. These distinct types result from various historical, social, and political-institutional developments (Artus 2023).

4. Literature Review on Transnational Care Workers in Germany

The Need for Elder Care

Since the proportion of older people in Germany is steadily increasing, it can be assumed that the number of people in need of long-term care will also rise. While in 2010, it was assumed that the number of people in need of long-term care would be around three million by 2023, it was already around 4.1 million in 2019 (Leiber and Österle 2022). According to the German Economic Institute, in 2023, it was already around five million (Enste and Anger 2023). The number of people in need of long-term care was severely underestimated. Because of the high demand of long-term care, a market for live-in care has developed since the 2000s for home-based care provided by domestic care workers (Leiber and Österle 2022). In Germany, about 3.3 million of the more than four million people in need of care, which is 80.2 percent, were already being cared for at home in live-in arrangements in 2019 (Enste and Anger 2023).

Furthermore, Leiber and Österle (2022) emphasize that there are limited official data on the number of people requiring care in 24-hour live-ins. This is because, according to the authors, domestic care work around the clock is an illegal market due to the 24-hour availability, and informal employment, meaning without or lacking social security protection and low wage (Tessier et al. 2022) mainly performed by transnational respectively migrant women. By the end of 2023, there were roughly about 985,000 live-in arrangements in Germany (Enste and Anger 2023). In 2017, it was estimated that 163,00 domestic workers were still working in live-in arrangements (Hielscher et al. 2017).

Transnational Care Workers and Intersectional Dynamics in Migration

In German private households with care needs, often times transnational care workers from Eastern Europe and women of color are employed (Habel and Tschenker 2022; Hielscher et al. 2017). They perform (un)paid domestic tasks such as cooking and cleaning, as well as caregiving tasks such as child-caring and the care of sick, elderly, or disabled people in institutional or private settings (Anderson 2002; Lutz 2002). According to estimates, there are around 600,000 care workers from Eastern Europe in Germany (Habel and Tschenker 2022). Transnational care workers typically migrate temporarily from regions such as East Africa, Southeast Asia, Eastern Europe, Mexico, the Caribbean, and South America to the United States or Germany, often maintaining connections with their families in their home countries. Most domestic care workers who migrate to Germany originally come from Central and Eastern European countries such as Bulgaria, Poland, Romania, Ukraine, and Hungary (Lutz 2011).

The intersectional analysis of transnational working and living conditions of care workers in private households is a broadly researched phenomenon in German gender and migration research (Kniejska 2016; Lutz 2007, 2011, 2016; Lutz and Palenga-Möllenbeck 2014, 2015; Theobold, 2024). Transnational migration is considered as both regular and irregular movement between two or more places, including a lifestyle that is designed to focus on multiple geographical and social spaces (Pries 2001a, 2001b). Transnational lifestyles among care workers can, according to Lutz (2007, 2016), be attributed to strict gender and migration regimes in welfare states like Germany, which do not allow for permanent settlement in the host country (Hess 2005). Migration policy restrictions of care workers from non-EU countries are strategically circumvented using the so-called rotation principle, where several care workers share employment in a German household for certain periods (Emunds 2016; Kniejska 2016). Studies by Karakayali (2010), Kniejska (2016), and Lutz (2011) document this informal care work through interviews with care workers in Germany and Central and Eastern Europe. In particular, Lutz examined the work and life situation of care workers who migrated from Eastern Europe to Germany. She did so from an intersectional perspective. Furthermore, she theoretically linked race, gender, and migration in the transnational space, thereby drawing attention to interrelated aspects of social and political inequality (Lutz, 2011; Lutz and Amelina 2017). However, she did not include organizing efforts in her research.

The above-mentioned studies have shown that domestic care workers face precarious working and living conditions in Germany. They are excluded from the social welfare system and from labor law. Studies also show racism in the workplace, especially against Eastern European care workers.

5. Approaches to Collectively Organize Transnational Care Workers

Union Organizing of Transnational Care Workers in Germany

The current state on the union organizing of transnational care workers in Germany can be summarized on the basis of a few academic publications: This includes Bomert's (2020) discourse analysis from social work on the lack of presence of transnational care workers' interests in public discourse. There is also a brief study and Schilliger and Schilling (2014) about a project of union (self-)organizing of Eastern European care workers in Switzerland. The researchers conclude that visible organizational structures for transnational care workers in private households are hardly present in Germany (Bomert 2020; Schilliger and Schilling 2014). An exception is the German Trade Union Federation's initiative Fair Mobility, which supports transnational care workers in home-based care arrangements. Since 2011, Fair Mobility has provided a contact point for care workers from Central and Eastern European countries at thirteen locations in Germany. Multilingual consultants with migration backgrounds support in labor and social law, accompany court proceedings (Federal Labor Court 2021, German: *Bundesarbeitsgericht*), do social media work, and lobby for a legal and safe employment environment in home-based care arrangements (Oblacewicz and Petö 2022). However, structural approaches for collective representation and union organizing are missing. Challenges for organizing include a lack of attention to care work as wage labor by trade unions (Ally 2005), isolation in private households, legal restrictions regarding residency, and an informal employment relationship due to a high identification with the ethos of care. These barriers challenge collective negotiations of employment contracts and strikes (Rudolph and Schmidt 2020). The German United Services Trade Union published a position paper on how to improve the working conditions of transnational care workers and their organizing potential. The brochure *Raus aus der Schwarzarbeit—Gute Arbeit in Privathaushalten* (translation by author: Exit Illegal Work—Decent Work in Private Households) focuses on labor market and employment policy approaches to regulate the employment relationship (Steffen 2015). One proposal is the establishment of service centers in municipalities as intermediary offices with counseling and qualification programs for employees such as transnational care workers. By collaborating with local stakeholders and organizations, these service centers can help to manage the regional demand for care workers in private households and promote the integration of transnational care workers into the local labor market (Steffen 2015). Thus, unions would have the opportunity to influence labor market policy processes. So far, these centers have not been established throughout the entire country. Additionally, these approaches by the German United Services Trade Union do not include an intersectional perspective regarding approaches to organizing transnational care workers. Restrictions with regards to migration and residency statuses such as tourist visas as well as racial and gendered aspects are not considered enough. However, addressing these issues is crucial when organizing transnational care workers and the challenges that they are facing.

Union Organizing of Transnational Care Workers in the United States

The current state, challenges, and organizing efforts of transnational care workers in the United States are similar to those in Germany. According to a study by the National Domestic Worker Alliance (NDWA), transnational care workers are affected by precarious working and living conditions and excluded from the social security system as well as labor protection laws (Burnham and Theodore 2012). Studies show racism in everyday employment, especially for Black care workers due to the historical connection of care work having been performed by formerly enslaved Black women, and later, Asian and South American women (Glenn 1992). Furthermore, Ehrenreich and Hochschild (2002) show the development of global care chains, and Parreñas (2005) comparatively studied transnational motherhood among Filipino care workers in Los Angeles and Rome through interviews. In the United States, union organizing of Asian, Mexican, Black, and precariously employed workers in so-called worker centers was promoted in the 1970s and 1980s following the social movements for civil rights in the 1960s (Fine 2006; Gordon 2005). Benz's (2014) research in labor sociology is noteworthy. Through expert interviews and observations, she examined the transformative character of worker centers between unions and social movements. For undocumented care workers, the DWU was founded as a multiethnic and transnational union in New York (Benz 2014; Guglielmo and Joffroy 2020; Norton et al. 1970).

Organizational Power Resources of Transnational Care Workers

The research for examining the union organizing of transnational care workers integrates practices of collective organizing (Bergmann 1979) and the power resource approach (Dörre and Schmalz 2014), considering the influence of welfare state models (Esping-Andersen 1990) on industrial relations. However, the power resources approach (Dörre and Schmalz 2014) provides the main theoretical framework. The theory addresses the question of what action possibilities and power resources unions possess for their transformation or renewal and what resources employees have for successfully pursuing their collective interests. There are four levels of power for employees: structural power, organizational power, institutional power, and societal power (Dörre and Schmalz 2014).

Structural power is the power to disrupt, which wage-dependent employees possess regardless of their access to collective representation. There is a distinction between production power and market power. Production power refers to the work stoppages by wage-dependent employees to enforce their interests (e.g., better working conditions) against capital. This includes circulation power (e.g., of workers in the transport sector on transport routes) and reproductive power. Care workers have reproductive power in kindergartens, care facilities, or private households. They can disrupt other wage-dependent workers from engaging in their employment, thereby influencing other areas of the economy. Market power, as the second form of structural power, is the influence employees have on the labor market through qualifications and job changes (Dörre and Schmalz 2014).

Organizational power arises from the unionization of worker organizing. Organizational power can be interpreted using metrics such as membership numbers (recruitment of new member groups through organizing and campaign work), infrastructure (office space and revenue), organizational efficiency (effective workflows, sensible use of resources), and member participation (volunteer union work and strikes) (Dörre and Schmalz 2014). The concept of "organizing" (Bremme et al. 2007; Dörre 2008; Dribbusch 2007; Schmalstieg and Choi 2009) is particularly discussed to recruit new member groups and increase membership numbers for more organizational efficiency in unions (Dörre and Schmalz 2014). Organizing aims to expand the action power of individuals and organizations beyond traditional forms of representation in unions, traditions, and parties into new social actors, including initiatives, networks, and social movements (Schmalstieg 2010; Rehder 2013). For unions, organizing transnational care workers can potentially mean recruiting a new member group, thereby strengthening membership numbers and organizational power. For transnational care workers, organizing approaches such as anti-racist and feminist self-empowerment and leadership development in union training seminars can be relevant to enforce their interests against employing families (Benz 2014). For the effective use of organizational power, a certain flexibility in organizational structure and action is needed. Flexibility can be achieved by organizing new member groups, such as transnational care workers, and considering the interests and resources of new members (Dörre and Schmalz 2014). Flexibility also involves overcoming traditional union action patterns, such as primarily addressing male occupational groups (Henninger 2017), or the assumption of unorganizability of care workers in private households and closing representation gaps in trade unions regarding women, particularly women of color (Ally 2005).

Institutional power is sustainable political influence, combining organizational power and structural power (Dörre and Schmalz 2014). The organizing of transnational care workers and their representation by unions can advance the influence on (inter)national labor, migration, and care policies at the political level and overcome the separation of economy and care in politics and society (Benz 2014; Winker 2015). This institutional power can be recognized and made visible by unions in Germany.

Societal power involves forming alliances with solidarity partners to pool forces. Societal power can draw on cooperation power or discourse power. The latter refers to influencing public discourses, such as recognizing care work in private households as paid work. Cooperation power strengthens organizational power through collaboration with social groups (churches, NGOs, students, scientists, leftists). While collaborating, the resources of solidarity actors can be used. For this, bridge builders are needed, who are involved in both union and non-union contexts. Through protests, initiatives, and transnational networks, political pressure can be amplified, as successfully done by the DWU with the Domestic Workers' Bill of Rights campaign (Benz 2014).

6. Conclusion

The paper highlights the need for the organizing and recognition of the power of transnational care workers from an intersectional perspective in Germany. This means that to effectively research how to unionize transnational care workers, it is essential to combine studies on transnational care workers from an intersectional perspective in gender studies with research on organizing care workers as employees in labor studies, ensuring that intersectionality is taken into account.

Furthermore, the paper shows the urgent need for effective union organizing of transnational care workers in Germany, inspired by successful strategies from the United States, particularly those implemented by the DWU. By integrating intersectionality as a key conceptual framework, the paper draws attention the precarious and irregular working conditions faced by transnational care workers and the significant challenges to their unionization in Germany. It emphasizes the importance of intersectional approaches in addressing complex social inequalities, drawing lessons from the United States more advanced intersectional organizing practices.

The working paper aimed to fill a notable gap in German labor sociology by examining the power resources of unions and care workers, despite their precarious conditions. By covering practices of collective action, organizing, and the power resource approach, the paper offered a comprehensive analysis of potential strategies for organizing transnational care workers in private households. The research underscores the critical role of structural, organizational, institutional, and societal power in unionization efforts and the need for flexible, inclusive union practices that embrace the diverse realities of transnational care workers.

In conclusion, this paper not only addressed a significant gap in the literature but also provided practical insights for enhancing the union organizing of transnational care workers in Germany. By learning from the successes of the DWU and other US-based initiatives, German unions might be able to develop more intersectional, inclusive, and effective strategies to improve the working conditions and representation of this vital yet marginalized workforce.

References

Ally, Shireen. 2005. "Caring about Care Workers. Organizing in the Female Shadow of Globalization." *Labour, Capital and Society*, Vol. 38, no. 1 & 2, pp. 185-207.

Anderson, Bridget. 2002. "Just Another Job? The Commodification of Domestic Labor. In Barbara Ehrenreich and Arlie R. Hochschild, ed., *Global woman: Nannies, maids and sex workers in the new economy*. London: Granta Publications, pp. 104-114.

Artus, Ingrid. 2023. "Industrielle Beziehungen in internationaler Perspektive." In Rainer Bohn, Hartmut Hirsch-Kreinsen, Sabine Pfeiffer, and Mascha Will-Zocholl, ed., *Lexikon der Arbeits- und Industriesoziologie*. Baden-Baden: Nomos, pp. 219-223.

Benz, Martina. 2014. *Zwischen Migration und Arbeit. Worker Centers und die Organisierung prekär und informell Beschäftigter in den USA*. Münster: Westfälisches Dampfboot.

Bergmann, Joachim. 1979. *Beiträge zur Soziologie der Gewerkschaften*. Berlin: Suhrkamp Verlag.

Bomert, Christiane. 2020. *Transnationale Care-Arbeiterinnen in der 24-Stunden-Betreuung. Zwischen öffentlicher (Un-)Sichtbarkeit und institutioneller (De-)Adressierung*. Wiesbaden: Springer.

Bremme, Peter, Fürniß, Ulrich, and Meinecke, Ulrike, eds. 2007. *Never work alone. Organizing—ein Zukunftsmodell für Gewerkschaften*. Hamburg: VSA

Bundesarbeitsgericht. 2021. "Mindestlohn–Bereitschaft–ausländische Betreuungskräfte." 5 AZR 505/20. https://www.bundesarbeitsgericht.de/entscheidung/5-azr-505-20

Burnham, Linda, and Theodore, Nik. 2012. "Home Economics. The Invisible and Unregulated World of Domestic Work." National Domestic Worker Alliance. https://workercenterlibrary.org/wp-content/uploads/2021/08/HomeEconomicsReport.pdf

Dribbusch, Heiner. 2007. "Das Organizing Modell. Entstehung, Varianten und Umsetzung." In Peter Bremme, Ulrike Fürniß, and Ulrich Meinecke, ed., *Never work alone. Organizing—ein Zukunftsmodell für Gewerkschaften*. Hamburg: VSA, pp. 24-52.

Dörre, Klaus. 2008. "Die strategische Wahl der Gewerkschaften. Erneuerung durch Organizing." *WSI-Mitteilungen*, Vol. 61, no. 1, pp. 3-10.

Dörre, Klaus. 2014. "Prekarität als Konzept kritischer Gesellschaftsanalysen—Zwischenbilanz und Ausblick. In Ethik und Gesellschaft." *Ökumenische Zeitschrift für Sozialethik*, Vol. 2, pp. 1-27.

Dörre, Klaus and Schmalz, Stefan. 2014. "Der Machtressourcenansatz. Ein Instrument zur Analyse gewerkschaftlichen Handlungsvermögen." *Industrielle Beziehungen*, Vol. 21, no. 3, pp. 217-237.

Barbara Ehrenreich, and Arlie R. Hochschild. 2002. *Global woman: Nannies, maids and sex workers in the new economy*. London: Granta Publications.

Emunds, Bernhard. 2016. "Menschenunwürdige Pflegearbeit in deutschen Privathaushalten. Sozialethische Bemerkungen zu den Arbeitsverhältnissen mittel und osteuropäischer Live-Ins." *Jahrbuch für Christliche Sozialwissenschaften*, Vol. 57, pp. 199-224.

Enste, Dominik, and Anger, Christina. 2023. "Haushaltshilfen: Nur selten angemeldet." *IW-Kurzbericht*, Vol. 76, pp. 1.3.

Esping-Andersen, Gösta. 1998. "Die drei Welten des Wohlfahrtskapitalismus. Zur Politischen Ökonomie des Wohlfahrtsstaates." In Stephan Lessenich, and Ilona Ostner, ed., *Welten des Wohlfahrtskapitalismus*. Frankfurt a.M., New York: Campus, pp. 19-56.

Fine, Janice. 2006. *Worker Centers. Organizing Communities at the Edge of the Dream*. Ithaca, New York: Cornell University Press.

Glenn, E. Nakano. 1992. "From Servitude to Service Work: Historical Continuities in the Racial Division of Paid Reproductive Labor." *Signs*, Vol. 18, no. 1, pp. 1-43.

Gordon, Jennifer. 2005. *Suburban Sweatshops. The Fight for Immigrant Rights*. Cambridge, MA: Belknap Press of Harvard University Press.

Guglielmo, Jennifer, and Joffroy, Michelle (2020). "A History of Domestic Work and Worker Organizing." https://www.dwherstories.com/timeline/radical-women-of-color-feminisms?prev=/timeline

Habel, Simone, and Tschenker, Theresa. 2022. "Organisation von Arbeitszeit in der Live-In-Pflege. Eine interdisziplinäre Untersuchung von sechs 'Pionieren' der Vermittlungsagenturen." https://www.boeckler.de/fpdf/HBS-008305/p_study_hbs_471.pdf

Henninger, Annette. 2017. "Geschlechterpolitische Verschiebungen in der deutschen Mindestlohndebatte: 'Equal pay' für Frauen oder für männliche Leiharbeiter?" *Industrielle Beziehungen*, Vol. 24, no. 2, pp. 135-155.

Hess, Sabine. 2005. *Globalisierte Hausarbeit*. Wiesbaden: VS Verlag für Sozialwissenschaften.

Hielscher, Volker, Kirchen-Peters, Sabine, and Nock, Lukas. 2017. "Pflege in den eigenen vier Wänden: Zeitaufwand und Kosten. Pflegebedürftige und ihre Angehörigen geben Auskunft." https://www.boeckler.de/pdf/p_study_hbs_363.pdf

Karakayali, Juliane. 2010. *Transnational Haushalten. Biographische Interviews mit "care workers" aus Osteuropa.* Wiesbaden: VS Verlag für Sozialwissenschaften.

Kniejska, Patrycja. 2016. *Migrant Care Workers aus Polen in der häuslichen Pflege. Zwischen familiärer Nähe und beruflicher Distanz.* Wiesbaden: VS Verlag für Sozialwissenschaften.

Leiber, Simone, and Österle, August. 2022. "Formalisierung des Informellen. Die Regulierung der '24-Stunden-Betreuung' in Österreich und Deutschland." *WSI-Mitteilungen*, Vol. 75, no. 5, pp. 379-385.

Lutz, Helma. 2002. "In fremden Diensten. Die neue Dienstmädchenfrage als Herausforderung für die Migrations- und Genderforschung." In Karin Gottschall and Birgit Pfau-Effinger, ed., *Zukunft der Arbeit und Geschlecht.* Wiesbaden: VS Verlang für Sozialwissenschaften., pp. 161-182.

Lutz, Helma. 2007. "'Die 24-Stunden Polin.' Eine intersektionelle Analyse transnationaler Dienstleistungen." In Cornelia Klinger, Gudrun-Axeli Knapp, and Birgit Sauer, ed., *Achsen der Ungleichheit. Zum Verhältnis von Klasse, Geschlecht und Ethnizität.* Frankfurt a. M., New York: Campus, pp. 210-234.

Lutz, Helma. 2011. *The New Maids: Transnational Women and the Care Economy*. London: Zed Books.

Lutz, Helma. 2016. "'Good Motherhood'—A Dilemma for Migrant Women from Eastern Europe." In Anna Amerlina, Horvath Kenneth, and Bruno Meeus, ed., *An Anthology of Migration and Social Transformation. European Perspectives*. Wiesbaden: VS Verlang für Sozialwissenschaften, pp. 245-258.

Lutz, Helma, and Amelina, Anna. 2017. *Gender, Migration, Transnationalisierung. Eine intersektionelle Einführung.* Bielefeld: Transcript.

Lutz, Helma and Palenga-Möllenbeck, Ewa 2014. "Care-Migrantinnen im geteilten Europa—Verbindungen und Widersprüche in einem transnationalen Raum." *Soziale Welt*, Vol. 20, pp. 217-231.

Lutz, Helma and Palenga-Möllenbeck, Ewa. 2015. "Care-Arbeit, Gender und Migration: Überlegungen zu einer Theorie transnationaler Migration im Haushaltssektor in Europa." In Uwe Meier-Gräwe, ed., *Die Arbeit des Alltags. Gesellschaftliche Organisation und Umverteilung.* Wiesbaden: VS Verlang für Sozialwissenschaften, pp. 181-200.

Norton, E. Holmes, Williams, Maxine, Beal, Frances, and La Rue, Linda. 1970. "Black Woman's Manifesto." https://repository.duke.edu/dc/wlmpc/wlmms01009

Oblacewicz, Justyna, and Petö, Bernadett. 2022. "Informationen zur Branche der 'Häuslichen Betreuung.' Erfahrungen aus der Beratungspraxis von Faire Mobilität." https://www.faire-mobilitaet.de/haeusliche-betreuung

Parreñas, R. Salazar. 2005. *Children of Global Migration: Transnational Families and Gendered Woes*. Stanford, CA: Stanford University Press.

Pries, Ludger. 2001a. *International Migration*. Bielefeld: Transcript.

Pries, Ludger. 2001b. "The Approach of Transnational Social Spaces: Responding to New Configurations of the Social and the Spatial." In Ludger Pries, ed., *New Transnational Social Spaces*. Routledge, pp. 3-37.

Rehder, Britta. 2013. "Vom Korporatismus zur Kampagne? Organizing als Strategie der gewerkschaftlichen Erneuerung." In Wolfgang Schroeder, ed., *Handbuch Gewerkschaften in Deutschland*. Springer VS, pp. 241-264.

Rudolph, Clarissa, and Schmidt, Katja 2019. *Interessenvertretung und Care. Voraussetzungen, Akteure und Handlungsebenen*. Münster: Westfälisches Dampfboot.

Rudolph, Clarissa, and Schmidt, Katja. 2020. "Vergeschlechtlichung und Interessenpolitik in Care-Berufen—das Beispiel Pflege." In Ingrid Artus, Nadja Bennewitz, Annette Henninger, Judith Holland, and Stefan Kerber-Clasen, ed., *Arbeitskonflikte sind Geschlechterkämpfe. Sozialwissenschaftliche und historische Perspektiven*. Westfälisches Dampfboot.

Schilliger, Sarah. 2013. "Care-Migration: Kampf der Hausarbeiterinnen um transnationale Wohlfahrt und Rechte." *Widerspruch: Beiträge zu sozialistischer Politik*, Vol. 62, no. 32, pp. 51-59.

Schilliger, Sarah. and Schilling, Katharina. 2014. "Care-Arbeit politieren: Herausforderungen der (Selbst-) Organisierung von migrantischen 24h-Betreuerinnen." *Femina Politica—Zeitschrift für feministische Politikwissenschaft*, Vol. 26, no. 2, pp. 101-116.

Schmalstieg, Catharina, and Choi, Hae-Lin. 2009. "Gewerkschaften und Prekarität. Neue Wege des Organizing." In Robert Castel, and Klaus Dörre, ed., *Prekarität, Abstieg, Ausgrenzung. Die soziale Frage am Beginn des 21. Jahrhunderts. Campus*, pp. 359-367.

Schmalstieg, Catharina. 2010. "Lernen von US-Gewerkschaften: Organizing." In Samuel Greef, S., Kalass, V., and Schroeder, W., ed., *Gewerkschaften und die Politik der Erneuerung—Und sie bewegen sich doch*. Düsseldorf: Hans-Böckler-Stiftung, pp. 207-226.

Schwenken, Helen 2005. "'Domestic Slavery' versus 'Worker Rights': Political Mobilization of Migrant Domestic Workers in the European Union." Working Paper 116. The Center for Comparative Immigration Studies.

Steffen, Margret. 2015. "… raus aus der Schwarzarbeit. Gute Arbeit in Privathaushalten. Europäische Erfahrungen und mögliche Gestaltungsansätze der Beschäftigung osteuropäischer Haushaltshilfen und Pflegekräfte." Berlin: ver.di.

Tessier, Lou, De Wulf, Nathalie, and Momose, Yuta. 2022. "Long-term Care in the Context of Population Ageing: A Rights-based Approach to Universal Coverage." ILO Working Paper 82. International Labor Organization. https://www.ilo.org/wcmsp5/groups/public/---ed_protect/---soc_sec/documents/publication/wcms_858784.pdf

Theobald, Hildegard. (forthcoming). "Migrant Care Workers in Home Care Settings in Germany: Inequality Dynamics and Policy Interactions." In Başak Akkan, Julia Hahmann, and Melanie Kuhn, ed., *Analysing Overlapping Inequalities in the Welfare State: Methodological Strengths and Challenges of Intersectional Approaches*. Springer Publishing House.

Winker, Gabriele. 2015. *Care Revolution. Schritte in eine solidarische Gesellschaft*. Bielefeld: Transcript Verlag.

X. Working in the Platform Economy

Personalized Work-Time Standards for Delivery Workers

MIN LI
NAN SUN
ZHENHUI BAI
South China University of Technology

Abstract

Work-time standards are a fundamental component of labor standards. However, with the rapid expansion of location-based digital labor platforms, determining work-time standards for gig delivery workers has become a pressing issue. Currently, delivery time standards on these platforms are algorithmically determined. The underlying logic of these standards still adheres to the principles of scientific management, which aim to define the time required by a qualified, capable, and trained operator working at a standard pace to complete a specific task. Digital labor platforms, however, fail to account for individual differences among workers, leading to numerous labor disputes. Gig delivery workers can register and begin or end work at any time. Their workplaces are open, and their work hours are fragmented. Compared to the closed or fixed workplaces and traditional 8-hour continuous work schedules, the work time and workspaces of gig delivery workers exhibit unique characteristics. This shift means that the assumptions underpinning work-time standard research have fundamentally changed. This study, from the perspective of China's digital development, uses big data to propose principles and calculation methods for a personalized work-time standard system. This system is designed to account for the individual circumstances of each delivery worker as well as real-time traffic conditions for each delivery task. The paper also discusses regulatory changes concerning work-time standards for delivery workers.

Introduction

Humanity has long strived to overcome the limitations of time and space, seeking to expand the finite dimensions of life, including work time and workspace. The development of digital labor platforms has introduced significant opportunities for change. With advancements in modern communication technology, the rise of digital labor platforms has become a major trend in the global labor market.

Through innovations in cloud computing, mobile internet, and data storage technologies, digital labor platforms have significantly enhanced the exchange of information, products, and services between consumers and providers. The International Labor Organization (ILO) defines digital labor platforms as systems that "allow individuals or business clients to arrange a ride, order food, or find a freelancer to develop a website or translate a document, among many other activities and assignments" (ILO 2021: 18).

Digital labor platforms can be categorized into two main types: online web-based platforms and location-based platforms. These platforms encompass a wide range of sectors, including micro-tasks, freelancing, competitions, medical consultations, travel, delivery, and household services (ILO 2021: 18).

The emergence of digital labor platforms has driven the rapid growth of the gig economy, becoming a global phenomenon (Berg et al. 2018), with China being no exception. According to the China Sharing Economy Development Report (2023), released by the National Information Center (2024), data show that in 2022, online takeaway revenue accounted for approximately 25.4% of the total revenue of China's catering industry, representing a 4-percentage-point increase compared to the previous year. From a user perspective, the penetration rate of online takeaway services among internet users reached 61.44%.

Meanwhile, the workforce in this sector continues to expand. For example, in 2022, more than 6.24 million delivery workers earned income through Meituan takeaway, with over 1 million active daily delivery workers (Meituan Research Institute 2023). In the first quarter of 2023, Meituan created 500,000 new positions for delivery workers and station managers. Additionally, Dada Group, in partnership with JD.com, generated hundreds of thousands of new jobs in 2022, recruiting delivery workers in over 2,700 counties and districts nationwide (National Information Center 2024).

These riders, formally known as Internet-based delivery workers, are included in the Occupational Classification Directory of the People's Republic of China (2022 edition). Their job description is summarized as "service workers who receive and inspect customer orders through location-based platforms and deliver ordered items to designated locations according to the order requirements" (Occupational Classification Directory Committee 2022).

Compared to traditional employment, most location-based delivery workers exhibit economic dependency rather than personal attachment to the platform. Their compensation is determined by a piece-rate system. In most Western countries, these workers are classified as independent contractors who complete tasks on demand and typically provide the necessary capital or equipment themselves (Stanford 2017).

In July 2021, the Ministry of Human Resources and Social Security of China, along with seven other departments, jointly issued the Guiding Opinions on Safeguarding the Labor Rights and Interests of Workers in New Forms of Employment (No. 56 of 2021 by the Ministry of Human Resources and Social Security). This document clarified the employment nature of workers who secure job opportunities through digital labor platforms. It stated: "If the establishment of a labor relationship is confirmed, the enterprise shall conclude a labor contract with the worker in accordance with the law," while "individuals conducting business activities independently on platforms or engaging in freelancing shall have their rights and obligations adjusted according to civil laws."

The document also introduced the concept of an incomplete confirmation of a labor relationship, signaling the emergence of a third type of labor relationship. This marked the beginning of an institutional framework distinct from both labor law and civil law. It represents an effort to explore the construction of a regulatory system for the era of a three-part division of labor and to establish standards for managing the third type of labor relationship. These developments highlight the need for further research on labor standards.

Work-time standards are a core component of international labor standards, directly influencing the establishment of key management systems such as minimum wage policies and social security frameworks. However, with changes as labor relationship regulations, evolve—from traditional fixed workplaces to boundaryless workspaces, and from fixed 8-hour workdays to fragmented, self-controlled work time, regulated schedules—the standard work times for workers in new forms of employment must be differentiated from those in traditional formal employment.

In the context of China's digital development, computational power—a core foundation of the digital economy—is entering a phase of rapid expansion. It is becoming a central productive force driving social and economic progress, ensuring interconnectedness and calculability across decision-making processes. This study has three primary objectives: first, to examine the scientific management principles and limitations underlying the establishment of work-time standards for digital labor platforms; second, to analyze the work-time and workspace characteristics of location-based delivery workers, assessing the technical conditions and feasibility of creating personalized work-time standards for them; and third, to propose strategies for developing personalized work-time systems alongside government regulatory recommendations.

Limitations on the Current Approaches for Work-Time Standards for Delivery Workers

Work-time study involves measuring work-time standards to determine the time required for a qualified, capable, and trained operator to perform a specific task at a normal pace under standard conditions (Yi and Guo 2022: 197). It is important to note that "work-time standard" has a specific meaning within time study. Based on the theory of the division of labor, workers are assigned to assembly lines where each individual repeatedly performs a specific task. Therefore, the work-time standard does not represent the total time for all processes involved in producing a workpiece but rather the time allocated for a single process (Fu 2003: 11).

Work-time standards enable the determination of the number of workers required at each station and the optimal production speed. This ensures the smooth and efficient operation of the assembly line, free of interruptions, facilitating mass production and reducing the cost per unit of product. Consequently, scientific management theory, as a system of methodology or industrial philosophy, marks a shift in organizational management practices. It moves away from the inefficiencies, conflicts, and disharmony inherent in traditional empiricism and personal authoritarian control. Instead, it governs industrial and labor relations during the Industrial Revolution with a spirit of scientific rigor.

However, research approaches to work-time standards based on Taylor's scientific management principles have two notable limitations. First, there is no objective standard for identifying qualified workers. Industrial engineers determine the normal pace through expert discussions, but currently recognized methods—such as time rating, predetermined time systems, and objective rating methods—still rely on subjective judgments. Engineers often develop impressions of the normal pace through training and select "qualified" workers based on these impressions. However, this approach fails to achieve a truly objective assessment.

Second, the research data used to establish work-time standards are derived from sampling methods. In *Principles of Scientific Management*, Taylor stated: "Select 10–15 different people, preferably from various enterprises and industries, who have special expertise in the job to be analyzed" (Taylor 1999: 81). This suggests that the number of worker samples in such studies is limited, and random biases in the data cannot be entirely avoided. Even in the absence of random bias, the sampling process itself may introduce systematic errors due to potential subjective intentions. As a result, sampled data may not accurately represent all workers in the sector. When coupled with the possibility of subjective bias in the sampling process, the resulting work-time standards may deviate to some extent from objective reality.

For food delivery platforms, such as Meituan in China, delivery times for workers are determined by the maximum value of four time parameters: the estimated arrival time from machine learning models; protection times based on city-specific geographies and traffic characteristics; protection times segmented by delivery stages (e.g., restaurant order preparation, rider arrival at the restaurant, rider travel, and delivery to the customer's location); and protection times based on delivery distance. Additionally, Meituan uses historical data to identify restaurants with consistently slow order preparation times and manually extends delivery times for food delivery workers in exceptional cases.

In 2021, Meituan Delivery publicly disclosed its algorithm rules for estimating delivery times, explaining that the algorithm considers factors such as city characteristics, the delivery process, and delivery distance. After calculating multiple estimated delivery durations, the algorithm selects the longest one as the final order delivery time to reduce pressure on workers. Furthermore, the platform has changed the estimated delivery time point to an estimated delivery time window, providing workers with a buffer.

This approach demonstrates that the underlying logic of delivery time standard regulations for food delivery platforms continues to reflect the standardized principles of scientific management.

Food delivery platforms do not adhere to strict regulations regarding the qualifications of workers during the selection process. According to information from the official website of Meituan Delivery, individuals need to meet only four conditions to register as location-based delivery workers: they must be physically healthy, aged between 18 and 56, possess a smartphone and an electric bicycle, and pass a safety test. This suggests that location-based delivery workers are only required to fulfill basic criteria, such as age and physical fitness, with no specific requirements for professional skills (Wang 2020).

In practice, however, location-based delivery workers must use information technology to perform their tasks, making digital skills essential for the job. Proficiency in operating food delivery apps and smart hardware devices, as well as understanding algorithmic rules for delivery tasks, is necessary to accept orders online and complete deliveries successfully. Furthermore, delivery workers face complex and ever-changing work environments, such as malls, roads, and residential communities, requiring them to coordinate with merchants, customers, and security personnel. This necessitates skills in communication, spatial planning, and emotional control.

The position of location-based delivery worker thus demands certain physical and professional competencies. For instance, the National Occupational Skill Standards for Location-based Delivery Workers (2021 edition), published by the Chinese government, outlines required attributes such as general intelligence, communication skills, computational abilities, spatial perception, and color vision. Additionally, research conducted by the Meituan Research Institute identifies key professional skills for location-based delivery workers, including interpersonal communication, emotional control, proactive learning, operation of production tools, handling emergencies and anomalies, digital skills, and real-time spatial-temporal planning.

The research also highlights a significant positive correlation between skill level and income. Higher professional skill levels enable delivery workers to complete more orders within a given timeframe, thereby increasing their earnings.

According to Taylor's scientific management theory, workers should avoid working at speeds detrimental to their health and instead maintain a suitable rhythm and pace, leading to greater happiness and avoiding overburdening themselves (Taylor 1999: 342). However, in efforts to improve the operational efficiency of location-based platforms and enhance customer experience, algorithmic technology automatically calculates estimated delivery times for food delivery workers after orders are accepted. This requires workers to complete deliveries within the specified timeframe. If they exceed it, they risk penalties from the platform or negative feedback from customers.

The estimated delivery times, however, fail to account for individual differences among workers. Real-time algorithm tracking, combined with work-time standardization, intensifies the time pressure experienced by some workers during their shifts (Gao et al. 2023). Prolonged exposure to high-intensity labor often results in physical and psychological strain (Liu et al. 2022). As a result, food delivery platforms have become battlegrounds where personalized vocational skills clash with standardized delivery requirements, posing significant challenges.

On the one hand, workers can participate in food delivery jobs as long as they meet basic requirements. On the other hand, digital technology does not eliminate the need for vocational skills in location-based delivery roles, leading to skill disparities among workers. Although work-time standards on food delivery platforms account for factors like geographical distance and restaurant preparation times, they remain rooted in standardized thinking derived from scientific management theory, overlooking individual worker differences.

By prioritizing delivery efficiency and excessively focusing on standardizing delivery times, these platforms often disregard the capacity of less-skilled workers. Under pressure from merchants and customers, delivery workers are more likely to engage in traffic violations, such as riding against traffic, speeding, running red lights, and other dangerous behaviors (Chen et al. 2021; Grunebaum 2024). Field investigations reveal that delivery times are continuously shrinking, with empirical research indicating that shorter average delivery durations are a key factor driving traffic violations among delivery workers (Wang et al. 2019).

Thus, the contradiction between standardized work-time requirements and varying vocational skills underscores the need for personalized work-time standards that align with individual capabilities. Big data technology can play a pivotal role in developing such personalized standards, accounting for the unique characteristics and differences among workers.

New Requirements and Technological Support

As the principles of scientific management continue to gain broader acceptance and deeper influence, the development of work-time standards for traditional employment has become highly advanced. However, the methodological framework for these standards still primarily relies on conventional tools such as stopwatches, which are insufficient for addressing the requirements of emerging employment forms. The core issue lies in the shift in foundational conditions for researching work-time standards—specifically, the fundamental changes in work time and workspace associated with emerging employment forms, which differ significantly from those of traditional employment (Barrow 2017).

Characteristics of Work Time and Workspace for Delivery Workers

During the process of industrialization, an intriguing shift occurred in the control of work time and workspace, following a U-shaped trend. The low end of this trend represents minimal control, while the high end represents extensive control. Before the era of mass production, labor processes typically took place in home-based workshops. For instance, prior to industrialization, the East India Company procured cotton from the Far East and distributed it to these workshops for spinning. Although agreements were established regarding delivery time, quality, and quantity, merchants had no control over the work processes or time management within the workshops. This lack of oversight led to increased transaction costs.

In 1991, Ronald Coase observed this phenomenon and explored the origins of firms in his paper The Nature of the Firm, posing the famous question: "Why do firms exist?" (Coase 2009: 37). While examining Adam Smith's concept of capitalists, Coase noted that one of the defining features of firms is their function as substitutes for the price mechanism. He explained that the entrepreneur was actively engaged in the division of labor within the firm, consciously planning and organizing its activities.

From home workshops to large-scale manufacturing factories, workers became fixed to production lines within planned and organized work schedules. As industrialization progressed from Industry 1.0 to Industry 3.0, all work processes were tightly controlled. This evolution brought increasing precision to the control of time and processes, moving from the bottom of the inverse U-shaped curve to the top, ultimately achieving a state of lean production.

Throughout this transformation, labor and capital engaged in prolonged negotiations. Beyond disputes over the distribution of interests, workers also sought to break free from the constraints of time and space.

However, as economic activities enter the digital era, workers have more choices in employment forms, one of which is choosing to descend along another curve from the top of the inverse U-shaped curve. This means they can opt for new forms of employment that allow them to complete work within their own scheduled time, freeing themselves from being entirely constrained by time. They have the freedom to determine the start and end of their work hours, as well as the duration of their work. At the same time, new forms of employment also expand the concept of workspace, giving workers more flexibility in their working environment.

For example, location-based delivery workers can register and start work at any time, and they can also end their work at any time. Their workplaces are open, and their working time is fragmented. Additionally, due to relatively low entry barriers, some location-based delivery workers even work part-time outside of regular employment, without the need for interviews or waiting for approval to start working freely. Compared to closed or fixed workplaces and the traditional 8-hour continuous work system, the work time and workspace of location-based delivery workers have entirely new characteristics.

New Characteristics of Work Time for Delivery Workers

The work hours of location-based delivery workers exhibit fragmented characteristics rather than fixed schedules. In traditional employment, work hours are typically set, such as from 9 a.m. to 5 p.m. However, location-based delivery workers are not restricted in this manner. They can accept delivery tasks based on their own schedules and available time slots, allowing them to work in the morning, afternoon, evening, or on weekends, which provides greater flexibility.

New Characteristics of Workspace for Delivery Workers

The workspace of location-based delivery workers differs significantly from traditional modes of work. Traditional work typically occurs within specific offices, factories, or stores, whereas the work of location-based delivery workers takes place across vast areas within cities or regions. They must ride or drive through city streets to transport goods from one location to another. This means that their workspaces are open and diverse, potentially requiring them to navigate various traffic conditions, weather challenges, and road difficulties. These characteristics of work processes pose challenges to traditional Taylorist systems and the principles of scientific management.

New Work-Time Standards Feasibly Provided by Digital Technology

Applying traditional methods to study work-time standards for location-based delivery workers presents two major challenges. First, achieving a standard state is not feasible. The standard state refers to the working method determined after work studies are conducted using principles of scientific management, including procedure analysis, operation analysis, and motion analysis. It encompasses standard equipment and tools, standard procedures, standard motions, standard machine operating speeds, and standard working environments. However, the work process of location-based delivery workers occurs within a social context and is influenced by various factors. Delivery routes are provided by digital labor platforms and may be affected by non-standard factors such as traffic conditions, temporary road closures, and equipment malfunctions. Additionally, the transportation used by location-based delivery workers is individually provided and cannot form a standard configuration, with its conditions varying due to individual differences. Due to these uncertainties, achieving the standard state required by traditional work-time standard formulation is challenging.

The second challenge is the difficulty of selecting and determining qualified workers. Taylor's scientific management emphasizes the selection of trained workers to ensure they meet the requirements of the work-time standard. However, location-based delivery workers are diverse, with varying skill levels, work styles, and even physiological conditions. Due to the diversity of their work environment and tasks, it is challenging to establish a universally qualified worker standard. Therefore, traditional selection and training methods may not be applicable in this context.

Clearly, digital labor platforms cannot ensure that every location-based delivery worker is a qualified worker. In reality, location-based delivery workers do not need to undergo a selection process to be employed; thus, there is no guarantee that they are suitable for delivery jobs or that their physiological condition is normal during the delivery process. Currently, aside from location-based delivery workers on platforms like Meituan's Zhuan Song, who undergo simple training, the delivery workers of another job called Zhong Bao typically do not receive training. While platforms control the timing for accepting orders and the delivery time, the delivery speed of each worker cannot be managed, often resulting in instances of running red lights or even traffic accidents.

Therefore, delivery workers on location-based digital labor platforms face significant challenges in establishing consistent work-time standards during their work processes. These processes cannot be standardized because they are distributed across society and influenced by numerous external factors. Moreover, their workspace is open and dynamic, making it impractical to standardize each delivery route. Therefore, we propose an innovative solution: personalized work-time standards. This approach offers real-time, individualized standards—delivered within seconds—tailored to each worker's specific context and the real-time traffic conditions for each delivery task.

The implementation of personalized work-time standards is foreseeable in the future, as China's digital economy development will generate big data. With the emergence of big data, innovative decision models and computational power will continue to evolve. From a policy perspective, the implementation of personalized work-time standards is feasible. The Chinese government has outlined the goal of accelerating digital economic development in the Overall Plan for the Construction of Digital China, emphasizing objectives such as empowering economic and social development comprehensively, strengthening and optimizing the digital economy, and achieving leading global levels of digital development by 2035. Based on the leadership of the Chinese government, this is a strategic goal that can be achieved. By 2035, it is expected that the penetration rate of online takeaway users among internet users will further increase from 61.44% in 2022 (State Information Center 2023). It is believed that the workforce of location-based delivery workers will correspondingly expand significantly, making personalized work-time standards even more relevant.

From a technological perspective, this proposal is also feasible. Overcoming the limitations of traditional employment work-time standards requires reliance on big data support, and the prerequisite for acquiring big data is the realization of the Internet of Things (IoT). In China, society is rapidly moving toward an intelligent world, with data experiencing explosive growth and intelligent technology advancing rapidly. Various models and algorithms based on powerful computing continue to emerge, with increasingly diverse application scenarios. This innovation is becoming more pronounced, surpassing traditional imagination. It is expected that everything will be interconnected, all applications will be modellable, and all decisions will be computable, relying on ubiquitous computing power to accelerate intelligent omnipresence. This will enable the continuous release and superposition of the potential of data in computation. As a flagship company in the technology needed for the development of the digital economy, Huawei is fully capable of supporting real-time computing for new employment models and personalized work-time standards.

According to the national strategy outlined in the Overall Plan for the Construction of Digital China and the outlook for future IoT development, the telecommunications industry, represented by Chinese companies, can provide the technological infrastructure needed to realize the IoT and computable decision-making. This capability makes it possible to establish the principles and methods for formulating personalized work-time standards.

Principles and Approaches of Work-Time Standards Formulation for Delivery Workers

Principles for Work-Time Standards Formulation

Since the widespread adoption of Taylorism in workplaces, its principles for formulating work-time standards have been criticized for neglecting the needs and rights of individuals. For instance, when studying workers' movements, ten principles are typically followed, including symmetrical hand movements, minimizing physical exertion, and placing tools and materials within easy reach. Therefore, in the process of formulating work-time standards for delivery workers, more attention needs to be paid to principles of fairness and freedom, fully respecting the personalized differences of workers and embodying principles such as the effective release of data elements, precision, inclusiveness, and convenience in the digital society, as well as global leadership in application innovation, highlighting Chinese characteristics while balancing humanistic care and fair efficiency. Consequently, when formulating personalized work-time standards for new forms of employment represented by gig economy delivery workers, the following principles can be considered.

Principle of Efficiency and Fairness

Balancing efficiency and fairness is crucial in formulating personalized work-time standards. This involves ensuring that work-time standards not only enhance productivity but also protect workers' rights and welfare. Delivery workers operating under similar working conditions should have comparable work-time standards that align with their individual capabilities. Factors such as task complexity, traffic and weather conditions, delivery experience, and the proper functioning of delivery tools must be considered. For instance, some delivery workers may illegally modify electric bicycles to increase speed and range, which can affect fairness and safety.

Principle of Freedom

Delivery workers should have the right to choose their own work hours and locations to better accommodate their personal lives and needs. Personalized work-time standards should provide flexibility, allowing workers to operate without rigid time constraints.

Principle of Humanistic Care

When developing personalized work-time standards, it is essential to prioritize the physical and psychological well-being of delivery workers. They should not be required to work at unsuitable times or in adverse conditions, as this could negatively impact their health.

Principle of Transparency and Understandability

Work-time standards should be clear, transparent, and easy to understand. Delivery workers should be able to comprehend how their working times are determined and access relevant information as needed. Encouraging delivery workers to participate in the formulation process and incorporating their feedback can enhance the acceptability and practicality of these standards.

Principle of Data-Driven Decision-Making

The development of personalized work-time standards should be informed by comprehensive data and analysis. This includes historical work performance, task difficulty, and delivery worker feedback to ensure the creation of more accurate and effective standards.

Principle of Legal Compliance

Personalized work-time standards must comply with legal regulations to protect the rights of delivery workers. Based on empirical research findings, it may be necessary to modify or innovate existing regulatory frameworks to align with evolving work environments.

Considering these six principles to achieve personalized work-time standards characterized by fairness, freedom, and humanistic care for delivery workers will help balance work efficiency and worker rights, contributing to a labor rights protection system within Chinese contexts.

Work-Time Standard Formulation Approach for Delivery Workers

Based on the six principles for personalized work-time standard formulation for delivery workers, the following formula may be used to calculate a personalized work-time standard. It can be adjusted and expanded according to specific needs:

Personalized Work-Time Standard = Basic Working Time +
Individual Difference Adjustment + Real-Time Road Condition Adjustment + Other Adjustments

Basic Working Time: This refers to the time value displayed on map applications, representing the time required to complete tasks under normal circumstances. It is based on real-time route analysis, such as the optimal route provided by Baidu Maps.

Individual Difference Adjustment: This adjustment accounts for individual differences among delivery workers, such as their system-recorded speed, efficiency, experience, and physical condition. It can be applied as a multiplier to increase or decrease the basic working time based on these factors.

Real-Time Road Condition Adjustment: This adjustment considers the impact of real-time road conditions, including traffic, road quality, and weather. Modifications can be made using a percentage or a specific time value based on current conditions.

Other Adjustments: Additional factors, such as task difficulty, task type, working time periods, and seasonal demands (e.g., during the rainy season), should also be considered. These adjustments can be added as needed and calculated based on actual circumstances.

Once the personalized work-time standard is established, the design of the working time management system can be considered. After working continuously for 4 hours, delivery workers are required to take a mandatory 20-minute off-line break. The total online time per day cannot exceed 8 hours (part-time delivery workers are limited to a maximum of 4 hours online per day). Working beyond 8 hours is classified as overtime, and the delivery fee for a single order during this period is calculated using the overtime pay coefficient stipulated by existing laws. The legal working time is set at 40 hours per week, with a maximum of 5 hours of work per day.

It is important to note that this is just a study framework. Specific calculation formulas and weights for each part should be determined based on actual conditions. Regular feedback from delivery workers, platforms, and experts should be solicited for modifications to ensure the acceptance and feasibility of the work-time standard. Additionally, historical big data and new data obtained from the IoT should be utilized to continuously adjust and improve this formula to accurately reflect real situations and meet government regulatory targets and requirements.

Work-Time Standard Regulation Proposal for Delivery Workers

Challenges

The first challenge is the lack of support for international organizational conventions. The ILO currently does not have specific conventions regarding personalized work-time standards. However, the ILO does address issues related to work-time standards and has developed a series of international conventions and recommendations concerning work time, working conditions, and labor rights. For instance, ILO Convention No. 1 on Hours of Work (Industry) and Convention No. 30 on Hours of Work (Commerce and Offices) cover work-time standards in the industrial sector and the commercial and office sectors, respectively, stipulating principles regarding hours of work, rest days, and overtime pay. While these conventions provide an international framework for work time, they do not encompass the specific concept of personalized work-time standards. Thus, formulating specific conventions regarding personalized work-time standards at the international level may require further research and discussion to accommodate the unique needs of different countries and industries.

The second challenge pertains to data usage permissions. Calculating personalized work-time standards requires big data and complex algorithms to consider individual differences, road conditions, task difficulty, and other factors. Relying solely on digital labor platforms for data collection may necessitate substantial technological and resource investments to establish and maintain these systems, significantly increasing platform operating costs. Additionally, as personalized work-time standards are proposed based on the future state of the IoT, the required big data is held by different institutions. Sharing this data would require corresponding laws and regulations to govern the collection, storage, and processing of personal data, thereby avoiding data privacy and national security issues.

The third challenge may arise from resistance among platform investors. Since devising the calculation methods for personalized work-time standards may be complex, digital labor platforms would need to invest more resources to ensure compliance with these standards. Platform operators may scrutinize this, potentially leading to increased costs and impacting business profitability.

The fourth challenge lies in the government's perspective. Digital labor platforms and governments need to continuously adjust and improve work-time standards to adapt to evolving work environments and demands. This necessitates the establishment of continuous monitoring and feedback mechanisms to ensure compliance with work-time standards by digital labor platforms and delivery workers. By leveraging the infrastructure developments, technology, and algorithms of the digital economy, the burden on the government can be significantly reduced. The key lies in transforming the government's regulatory system to keep pace with the advancement of the digital economy while maintaining close cooperation and negotiation with digital labor platforms and labor representative organizations.

Proposed Regulations

Implementing a personalized work-time standard system for delivery workers involves government oversight and regulation to ensure labor rights, safety, fair competition, and legal compliance. The government can adopt new regulatory measures.

First, it is necessary to establish new laws or amendments to existing laws and regulations. Clearly define the work-time standard system for digital labor platforms and the delivery industry, including methods for calculating personalized work time, minimum rest periods, overtime, and wage standards. Ensure delivery workers have sufficient rest time to avoid overexertion and enhance safety. Determine maximum daily and weekly working hour limits, as well as restrictions on consecutive work time. Set minimum wage levels and require platforms to pay these wages or provide appropriate benefits to safeguard the economic rights of delivery workers, including mechanisms for considering overtime or other circumstances.

Second, it needs to clarify data usage permissions and transparency. Digital labor platforms should be required to provide transparent information about work time and income, enabling delivery workers to understand how their work time and salaries are calculated. The government can mandate platforms to provide data to supervise and verify compliance with work-time standards.

Third, it should ensure work safety and provide social security. Digital labor platforms should be required to provide adequate insurance coverage for delivery workers' work-related accidents and medical expenses, ensuring compliance with safety regulations to reduce accident and injury risks. Additionally, it is essential to ensure that delivery workers are not discriminated against and can access work and treatment equally, regardless of gender, race, religion, or other factors.

Finally, it is important to establish penalty and supervision mechanisms. This involves formulating arbitration and penalty measures to address platform violations of work-time standards. In terms of labor dispute resolution, regulatory agencies or committees must be established to supervise and enforce compliance with work-time standards. Moreover, it is essential to encourage platforms and labor representative organizations to engage in collective bargaining on work-time standards to ensure that all parties' opinions are fully considered. Additionally, it is necessary to solicit feedback and opinions from delivery workers to continuously improve work-time standards and adapt to the rapidly changing digital economy.

Acknowledgments

This research is funded by Chinese National Social Science Foundation (grant number: 21BGL027).

References

Barrow, I. 2017. *The East India Company, 1600–1858: A Short History with Documents.* Hackett Publishing.

Berg, J., et al. 2018. *Digital Labor Platforms and the Future of Work—Towards Decent Work in the Online World.* International Labour Organization.

Chen, L., and P. Sun. 2021. *Super Mobility, Accelerated Circulation, and the Movement Away from Center: Reflections on the Labor of Mobility as Livelihood on Internet Platforms*. China Youth Study (in Chinese).

Coase R.H. 2010. *The Firm, The Market, and The Law*. Shanghai People's Press (in Chinese).

Fu, W.X. 2003. *Work time standard Development and Work Improvement*. Fuyou Modern Practical Enterprise Management Series. Xiamen University Press (in Chinese).

Gao, X.Y., et al. 2023. "Gig Worker's Work Pressure: Formation Mechanism and Scale Development." *Nankai Business Review* 26 (3,): 244–58 (in Chinese).

Grunebaum, D. 2024. "Malaysian Road Safety Institute Pushes for Better Training of Food Delivery Workers." VOA. https://www.voanews.com/a/economy-business_malaysian-road-safety-institute-pushes-better-training-food-delivery-riders/6195935.html

International Labour Organization (ILO). 2021. *World Employment and Social Outlook: The Role of Digital Labor Platforms in Transforming the World of Work.*

Liu, S. S., et al. 2022. "Algorithmic Management of Online Labor Platforms: Theoretical Exploration and Research Prospect." *Journal of Management World* 38 (2): 225–39 + 14–16 (in Chinese).

Meituan Research Institute. 2022. *The Report on Corporate Social Responsibilities* (in Chinese).

National Occupational Classification Compendium Revision Working Committee. 2022. *Occupational Classification Compendium of the People's Republic of China*. China Labor and Social Security Publishing (in Chinese).

Stanford, J. 2017. "The Resurgence of Gig work: Historical and Theoretical Perspectives." *Economic and Labor Relations Review* 28 (3): 382–401.

State Information Center. 2023. *China Shared Economy Report*. https://zjic.zj.gov.cn/ywdh/jjyx/202302/P020230224354330850066.pdf

Taylor, F.W. 1999. *The Principles of Science Management*. Tuanjie Press (in Chinese).

Wang, X. 2020. "Gig Economy Skills: The Key to 'Freedom of Choice' for Workers." *Exploration and Free Views*, No. 7: 29–31 (in Chinese).

Wang, Y.K., et al. 2019. "Analysis on Influencing Factors of Public Traffic Violation Behavior in Takeout Delivery." *Journal of Safety Science and Technology* 15, (12): 169-–74 (in Chinese).

Yi, S.P., and F. Guo. 2022. *Fundamentals of Industrial Engineering*. China Machine Press (in Chinese).

XI. Intersectionality and Worker Voice

A Literature Review of Intersectionality with Implications for Employment Discrimination Research and Practice

HELEN LAVAN
DePaul University

Abstract

This research seeks to answer the question of how the role of intersectionality has been researched from the perspective of employee relations, especially employment discrimination. ProQuest was used to identify 978 usable articles from 2014 through 2023 on the topics of intersectionality, work, and discrimination. The main findings include, in line with Critical Race Theory, that the terms "women" and "Black" were the most frequently occurring. The information does not include employee relations or industrial relations. While the term "union" was frequent, it is not clear that bargaining or negotiation by a union was involved. The current research does not reflect the reality of discrimination, in that employers comprising large segments of the labor force, such as manufacturing, were underrepresented. Instead, it is a dataset relating to family, public, and education. Recommendations on how to further develop theory and, ultimately, practice relating to intersectionality and employment discrimination are made.

Introduction

This research seeks to determine how the role of intersectionality has been researched from the perspective of employee relations, especially employment discrimination.

Intersectionality connects social categories, such as gender, race, class, sexuality, and ability. Some of these are afforded legal protections, while others are not. Some of these factors shape an individual's experiences and opportunities. Intersectionality recognizes that different forms of discrimination often intersect.

Intersectionality in employment discrimination has been given increased legal cognizance due to a recent position paper by the Equal Employment Opportunity Commission (EEOC) (Coleman et al. 2021). Recently, Pappoe (2019) noted that progress in resolving discrimination could be made by incorporating an intersectional lens in the design and implementation of laws and policies.

One aspect is a dearth of research intersectionality in the context of work and employment, which dearth has been noted for several years (McBride et al. 2015; Özbilgin et al. 2011; Sawyer et al. 2013). There have been calls from other authors to increase this research from an intersectional perspective (Bailey et al. 2019; Koellen 2021; Yadav and Lenka 2022). It is not a matter that there have not been recent literature reviews on intersectionality. They are not explicitly directed at employee relations (Garcia and Zajicek 2021; Koellen 2021; Nichols and Stahl 2019; Yadav and Lenka 2022). Emerging societal trends, including #MeToo, #TimesUp, #BlackLivesMatter, LGBTQ rights, and immigrant rights, may have changed the applicability of the earlier research.

Literature Review

The starting point of intersectionality might have begun with Ruggs et al. (2013). Following up on their approach were Sawyer et al. (2013), who noted that studying individual (marginalized) identities is good, but examining intersectionality is better.

Özbilgin et al. (2011) were early contributors to the debate on intersectionality, noting that the why's and how's of diversity other than gender also manifest as salient causes of the difference in experiences of the work–life interface.

In an introduction to a special issue on intersectionality published in *Gender in Management*, Corlett and Mavin (2014) noted concerns about intersectionality research. Their concern was about the framework and theory that could be shared in the study of intersectionality and identity. While acknowledging Critical Race Theory in feminism from the 1980s, McBride et al. (2015) argued that the field would benefit from an intersectional approach in work and employment relations studies. Al-Faham et al. (2019) concluded that "despite different views on conceptualization, application, and implications, intersectionality may nevertheless open new avenues of inquiry for scholars as well as opportunities for transformative coalition building in social movements."

The contribution of Bauer et al. (2021) was that intersectionality can be studied using quantitative and more traditional qualitative methods. Koellen (2021) and Yadav and Lenka (2022) focused on intersectionality from a diversity perspective. Both studies provided critical reviews and research directions for intersectional perspectives, an important approach to understanding diversity.

Recently, advocates for intersectionality approaches have been made. Salter et al. (2021) examined how intersectionality impacts work-related attitudes, and Handl et al. (2022) advocated for an intersectionality approach beyond gender.

Intersectionality Research by Sector

There are a variety of recent studies by sector, including healthcare, the public sector, and education.

Healthcare

Healthcare is probably the most developed of the sectors when considering intersectionality. Abrams et al. (2020) advocated for using intersectionality in qualitative health research. Similarly, Bauer et al. (2021), Kelly et al. (2021), and Richman and Zucker (2019) advocated using intersectionality to attain equity in healthcare.

Public Sector

Public sector intersectionality research includes the work of Healy et al. (2011), who advocated for employing intersectionality in qualitative health research. Nelson and Piatak's (2021) focus was on racially underrepresented minorities in the federal government. The study sample by LaVan and Martin (2024) was similar. Still, they delved more deeply into intersectionality, considering the mutability of the demographic characteristics.

Education

Nichols and Stahl (2019) conducted an interesting literature review on intersectionality in higher education. Earlier, Sanchez-Peña et al. (2016) examined the persistence of Latina faculty, considering their race, gender, and class.

Qualitative Methods and Grounded Theory

Although intersectionality research can be quantitative, recent authors have published advice on how to research intersectionality using qualitative methods (Bailey et al. 2019; Belotto 2018; Williams and Moser 2019). Some emphasis is being given to developing intersectionality through grounded theory methods (Charmaz and Thornberg 2021).

The main finding of the current study is that insufficient attention has been given to what comprises the existing research about intersectionality, discrimination, and work. At least a decade has passed since Ruggs et al. (2013) suggested intersectionality on marginalized populations on which to conduct discrimination research. The current research inventories the existing literature using the broadest possible search terms to discern what research on intersectionality and employment from a workplace setting exists and how future research should be focused.

Methodology

Research Questions

The main research question is how intersectionality is researched in the articles that comprise the database. Questions include the most frequent demographic, process, and research terms overall. What protected demographic bases are included in the literature? What are the contexts in which the studies took place? What research processes are used? How are employees/industrial relations involved? This research will enable a deeper understanding of intersectionality and future directions for the research and practice of intersectionality in employee relations.

Sample

Using ProQuest, the following search strategy was used to develop a sample of 978 articles from 2014 through 2023, with a word count of 5,700,000. Terms in the search strategy include intersectionality, work, and discrimination to discern content relating to intersectionality and employee relations.

Using NVivo content analysis, the following can be inventoried within the literature: (1) intersectional demographics, such as race, gender, age, health, etc.; (2) context variables, including workplace, sector (public, private), and healthcare; (3) theoretical backgrounds; (4) methodology; and (5) mutability of demographic characteristics. A total of 47 terms were analyzed. Some were considered nonrelevant and were excluded.

Results

Table 1 shows a selection of frequent terms that appeared in the top 500 terms in the database. It is not surprising, given the emphasis on critical race theory over time, that "women" and "Black" are the top two terms, followed closely by "discrimination." Another conclusion that can be drawn from this table is that "transgender" and "Asian" appeared more frequently than "disability" and "Hispanic," even though people with disabilities and people of Hispanic origin are more prevalent in the population.

The conclusions illustrated in Figure 1 are similar to those of Table 1: very few mutable characteristics were noted, except for pregnancy and mental illness, which have low frequency. Specifically, pregnancy is not permanent, and people with mental illnesses can be cured or at least treated. However, mutable characteristics were not frequently contained in the articles in the database.

As shown in Figure 2, the terms with the highest frequency are "family," "public," and "education." While "workplace" and "hospital" had moderate frequencies, "manufacturing" appeared in only 44 of the 978 articles. This is also disproportionate to the reality in which innumerable manufacturing employers exist. One conclusion is that prior intersectionality studies were not about the workplace but about family, public, and education.

Figure 3 shows a combination of research terms and theory terms. As a researcher of employment discrimination for at least 30 years, I expected "microaggression," "immutability," and "implicit bias" to appear far more frequently in this database. The two theories examined were critical race theory, with a frequency of 138, and grounded theory, with a frequency of 127.

Regarding methodology, one can conclude that "interviews" and "surveys" were approximately equal in frequency. "Qualitative analysis" was mentioned in 170 of the articles. This is unsurprising because qualitative analysis has long been accepted in healthcare research. At the bottom of the table, it can be seen that 613 of the articles mentioned "legal," but only 8 refer to "legal scholarship." Although it is recognized that not all of the research was conducted in the United States, the frequency of "EEOC" was less than expected.

Figure 4 shows the frequency of "diversity" (748) and "discrimination" (816). While "union" was mentioned in 184 of the articles, the union was probably not involved in bargaining or negotiating. The terms "employee relations" and "industrial relations" were mentioned in a few instances, mainly in the titles of journals, leading to the conclusion that the focus was not on the union's role in intersectionality research.

Table 1. Frequent terms in the top 500

Terms of Interest (Descending)	Demographic	Process	Research Terms
Women	*		
Black	*		
Discrimination		*	
Sexual	*		
White	*		
Race	*		
Education	*		
Family	*		
Psychology		*	
Racism		*	
Female	*		
School		*	
Stigma		*	
Transgender	*		
Asian	*		
Diversity		*	
Ethnicity		*	
Qualitative			*
Workplace		*	
Bisexual	*		
Lesbian	*		
LGBTQ	*		
Economic		*	
Language	*		
Muslim	*		
Sexuality	*		
Stereotypes		*	
Depression	*		
Old	*		
Religious	*		
Disabilities	*		
Hispanic	*		
Immigrants	*		

Figure 1. Demographic characteristics

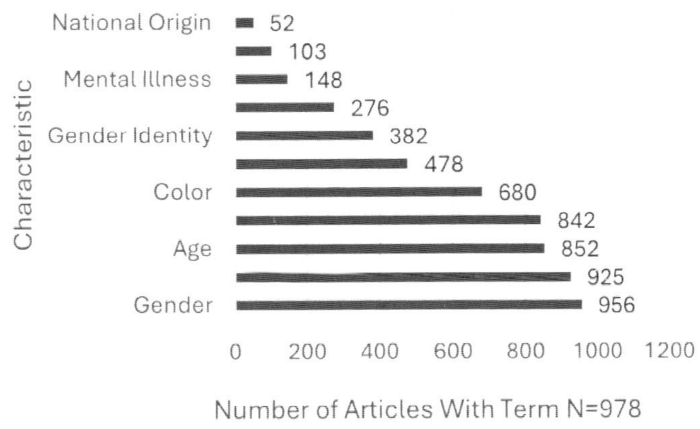

Figure 2. Context of articles (N = 978)

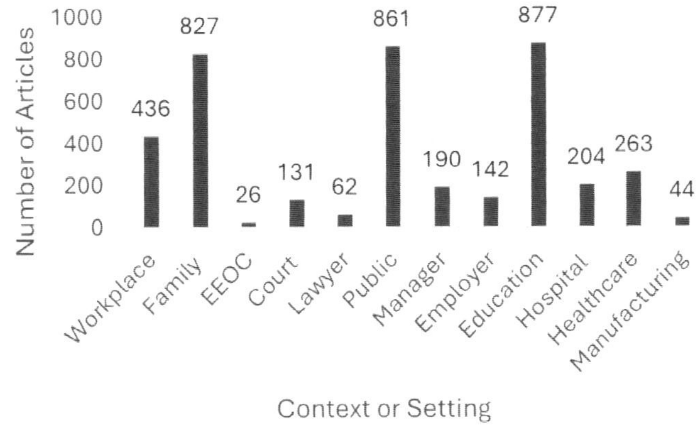

Figure 3. Theory and research terms

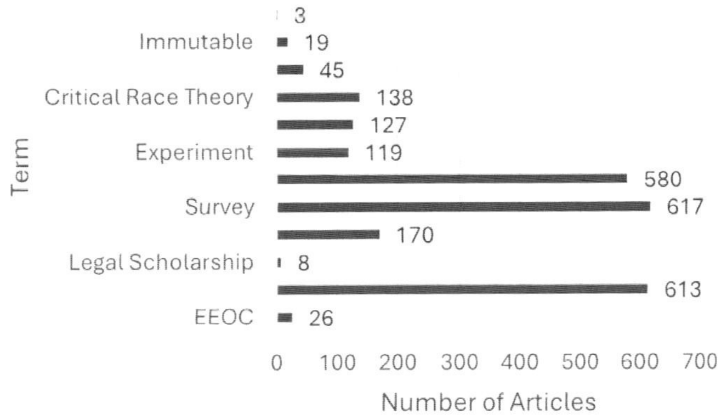

Figure 4. Employee relations terms

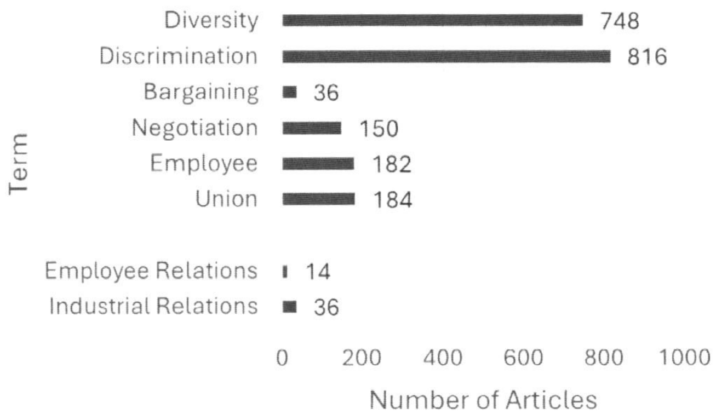

Direction for Further Research

Some activities that will enhance this study:

- Broaden the search strategy to include political science and law. Query HeinOnline to discern whether there are published studies in the legal literature.
- Separating earlier publications from more recent ones might also be desirable.
- This dataset might need to be cleaned in a process called decoding to discern whether there are homonyms that might distort the outcome. For example, the term "Black" could also be a surname. However, given the frequency of the term "Black" in this data set, it is not likely that it would distort the findings.
- Autocoding of the results could identify theoretical concepts that underlie most of the articles in the data set that are not identifiable by manual searching.
- Explicitly identify the industrial relations articles that comprise the database that was not identified by using the search term "industrial relations." One approach might be to examine journals related to business and society. It is recommended that the Australian Business Deans Council list be used (https://abdc.edu.au/abdc-journal-quality-list).

Some activities that will push the field forward:

- Researchers of employee relations should consider the contribution that qualitative research methods can make to understanding issues in employee relations.
- Editors of prominent employee relations and industrial relations journals call for special issues requiring interdisciplinary content on workplace intersectionality from an intersectionality perspective. Journal editors should be more receptive to manuscripts using qualitative data and grounded theory.
- While valuing all methodologies, use existing databases and real subjects who have been discriminated against. Some of these databases are widely available at low or no cost.
- Encourage legality, collective bargaining, and DEI as focal content.
- Enhance the study of poorly researched variables, including salient and mutable/immutable distinctions.
- Focus on major occupational groups with broad-ranging generalizability, not convenience samples.

References

Abrams, J.A., Tabaac, A., Jung, S., and Else-Quest, N.M. 2020. "Considerations for Employing Intersectionality in Qualitative Health Research." *Social Science & Medicine* 258: 113–38. doi: 0.1016/j.socscimed.2020.113138

Al-Faham, H., Davis, A.M., and Ernst, R. 2019. "Intersectionality: From Theory to Practice." *Annual Review of Law and Social Science* 15: 247–65. doi: 10.1146/annurev-lawsocsci-101518-042942

Bailey, J., Steeves, V., Burkell, J., Shade, L. R., Ruparelia, R., and Regan, P. 2019. "Getting at Equality: Research Methods Informed by the Lessons of Intersectionality." *International Journal of Qualitative Methods* 18: 1609406919846753. doi: 0.1177/1609406919846753

Bauer, G.R., Churchill, S.M., Mahendran, M., Walwyn, C., Lizotte, D., and Villa-Rueda, A.A. 2021. "Intersectionality in Quantitative Research: A Systematic Review of Its Emergence and Applications of Theory and Methods." *SSM - Population Health* 14: 100798. doi: 10.1016/j.ssmph.2021.100798

Belotto, M.J. 2018. "Data Analysis Methods for Qualitative Research: Managing the Challenges of Coding, Interrater Reliability, and Thematic Analysis." *Qualitative Report* 23 (11) 2622–33. doi: 10.46743/2160-3715/2018.3492

Charmaz, K., and Thornberg, R. 2021. "The Pursuit of Quality in Grounded Theory." *Qualitative Research in Psychology* 18 (3): 305–27. doi: 10.1080/14780887.2020.1780357

Coleman, A., Dupont, R., and Rivera, N. 2021. *In Pursuit of Pay Equity: Examining Barriers to Equal Pay, Intersectional Discrimination Theory, and Recent Pay Equity Initiatives.* US Equal Opportunity Employment Commission. https://www.eeoc.gov/pursuit-pay-equity-examining-barriers-equal-pay-intersectional-discrimination-theory-and-recent-pay

Corlett, S., and Mavin, S. 2014. "Intersectionality, Identity and Identity Work: Shared Tenets and Future Research Agendas for Gender and Identity Studies." *Gender in Management: An International Journal* 29 (5): 258–76. doi: 10.1108/GM-12-2013-0138

Garcia, T.C., and Zajicek, A. 2021. "Incorporating Intersectionality in Public Policy: A Systematic Literature Review. *"Humanity & Society* 46 (2): 271–90. doi: 10.1177/0160597620988591

Handl, M.N., Seck, S.L., and Simons, P. 2022). "Gender and Intersectionality in Business and Human Rights Scholarship." *Business and Human Rights Journal* 7 (2): 201–25. doi: 017/bhj.2022.12

Healy, G., Bradley, H., and Forson, C. 2011. "Intersectional Sensibilities in Analysing Inequality Regimes in Public Sector Organizations." *Gender, Work & Organization* 18 (5): 467–87. doi: 10.1111/j.1468-0432.2011.00557.x

Kelly, C., Kasperavicius, D., Duncan, D., Etherington, C., Giangregorio, L., Presseau, J., Sibley, K.M., and Straus, S. 2021. "'Doing' or 'Using' Intersectionality? Opportunities and Challenges in Incorporating Intersectionality into Knowledge Translation Theory and Practice." *International Journal for Equity in Health* 20 (1), Article No. 187. doi: 10.1186/s12939-021-01509-z

Koellen, T. 2021. "Diversity Management: A Critical Review and Agenda for the Future." *Journal of Management Inquiry* 30 (3): 259–72. doi: 10.1177/1056492619868025

LaVan, H., and Martin, M. 2024. "Inequality, Mutability and Intersectionality Differences in Discrimination in the Us Federal Public Employment Sector." *Employee Responsibilities and Rights Journal* 36 (1): 101–27. doi: 10.1007/s10672-023-09437-6

McBride, A., Hebson, G., and Holgate, J. 2015. "Intersectionality: Are We Taking Enough Notice in the Field of Work and Employment Relations?" *Work, Employment and Society* 29 (2): 331–41. doi: 10.1177/0950017014538337

Nelson, A., and Piatak, J. 2021. "Intersectionality, Leadership, and Inclusion: How Do Racially Underrepresented Women Fare in the Federal Government?" *Review of Public Personnel Administration* 41 (2): 294–318 doi: 10.1177/0734371x19881681

Nichols, S., and Stahl, G. 2019. "Intersectionality in Higher Education Research: A Systematic Literature Review." *Higher Education Research & Development* 38 (6): 1255–68. doi: 10.1080/07294360.2019.1638348

Özbilgin, M.F., Beauregard, T.A., Tatli, A. and Bell, M.P. 2011. "Work–Life, Diversity and Intersectionality: A Critical Review and Research Agenda." *International Journal of Management Reviews* 13 (2) 177–98. doi: 10.1111/j.1468-2370.2010.00291.x

Pappoe, Y.N. 2019. "The Shortcomings of Title VII for the Black Female Plaintiff." *University of Pennsylvania Journal of Law and Social Change* 22: 1-24. https://heinonline.org/HOL/P?h=hein.journals/hybrid22&i=7

Richman, L.S., and Zucker, A.N. 2019. "Quantifying Intersectionality: An Important Advancement for Health Inequality Research." *Social Science & Medicine* 226: 246–48 doi: 10.1016/j.socscimed.2019.01.036

Ruggs, E.N., Hebl, M.R., Law, C., Cox, C.B., Roehling, M.V., and Wiener, R. L. 2013. "Gone Fishing: I–O Psychologists' Missed Opportunities to Understand Marginalized Employees' Experiences with Discrimination." *Industrial and Organizational Psychology* 6 (1): 39–60. doi: 10.1111/iops.12007

Salter, N.P., Sawyer, K., and Gebhardt, S.T. 2021. "How Does Intersectionality Impact Work Attitudes? The Effect of Layered Group Memberships in a Field Sample." *Journal of Business and Psychology* 36 (6): 1035–52 doi: 10.1007/s10869-020-09718-z

Sanchez-Peña, M., Main, J., Sambamurthy, N., Cox, M., and McGee, E. 2016. "The Factors Affecting the Persistence of Latina Faculty: A Literature Review Using the Intersectionality of Race, Gender, and Class." *Proceedings of the 2016 IEEE Frontiers in Education Conference (FIE)*, Erie, PA.

Sawyer, K., Salter, N., and Thoroughgood, C. 2013. "Studying Individual Identities Is Good, but Examining Intersectionality Is Better." *Industrial and Organizational Psychology* 6 (1): 80–84. doi: 10.1111/iops.12012

Williams, M., and Moser, T. 2019. "The Art of Coding and Thematic Exploration in Qualitative Research." *International Management Review* 15 (1): 45–55. http://www.imrjournal.org/uploads/1/4/2/8/14286482/imr-v15n1art4.pdf

Yadav, S., and Lenka, U. 2022. "A Systematic Review of Job-Related Diversity and Future Research Directions." *Equality, Diversity and Inclusion: An International Journal* 41 (8): 1181–1208. doi: 10.1108/edi-12-2021-0324

XI. Intersectionality and Worker Voice

Unraveling the Journey of Women of Color in Corporate America

BRANDI R. MUÑOZ
Abilene Christian University

Abstract

This research paper explores strategies to enhance diversity, equity, inclusion, and belonging (DEIB) initiatives in organizational leadership, particularly for women of color (WOC) in corporate America. Despite their significant contributions, WOC face systemic barriers and underrepresentation in leadership roles. The findings highlight multifaceted barriers related to gender, race, and intersectional identities that span from pre-women's suffrage through the present that hinder WOC's access to leadership opportunities and career advancement. Key recommendations underscore the importance of targeted DEIB initiatives, such as mentorship programs, flexible work arrangements, and recognition of transformational leadership behaviors. The study contributes to the growing body of literature on DEIB by offering actionable strategies to promote equity and empowerment for WOC. This research emphasizes the critical need for organizational leaders to implement inclusive practices that support the professional growth and leadership development of WOC, thereby fostering a more equitable and diverse workplace.

Introduction

Qualitative studies reveal a disparity in the career progression of women, particularly women of color (WOC), compared to men or White women. WOC face unequal pay and limited career development opportunities (Lee and Wessel 2022; Oleschuk 2020). Despite this, there is a lack of research on the impact of disruptive changes in the US workplace that consider cross-cultural perspectives encompassing gender, race, age, socioeconomic status, and sexual orientation (Bawany 2022; Everly and Athey 2023; Pilcher and Whelehan 2004; Potter 2021; Shim, Dik, and Banning 2022; Smith 2000). The absence of such studies poses a threat to the advancement of WOC in the American workplace, especially in the aftermath of disruptive events like the 2020 COVID-19 pandemic (Acker 2006; Bawany 2022; Crichlow, Coburn, and Joseph 2020; Cubrich et al. 2022; Kim and Gege 2021; Landivar, Ruppanner, Scarborough, and Collins 2020; McClain 2020; O'Conner 1996; Oleschuk 2020; Phipps and Prieto 2020; Randsley de Moura et al. 2018; Smith 2000; Snyder 2008). Moreover, persistent challenges hinder WOC's socioeconomic growth, impeding their career advancement (Oleschuk 2020). Consequently, the proven positive contributions of WOC to corporate America, especially during disruptive changes, are jeopardized (Acker 2006; Bawany 2022; Bloch, Taylor, Church, and Buck 2020; Carli 2020; Diehl and Dzubinski 2016; Everly and Athey 2023; Fitzsimmons and Callan 2016; Glass and Cook 2018; Hoyt and Murphy 2016; Kaiser and Wallace 2016; Potter 2021; Simmons and Yawson 2022).

Hence, it is imperative to delve deeper into understanding how organizational leaders can optimize their diverse workforce, with a particular emphasis on career development for WOC (Barak 2016; Cinamon 2006; Ellemers 2014; Gray and O'Brien 2007; Hideg and Shen 2019; Lee and Wessel 2022). Without this crucial exploration, the consequences are profound (Acker 2006; Bawany 2022; Bloch, Taylor, Church, and Buck 2020; Carli 2020; Diehl and Dzubinski 2016; Fitzsimmons and Callan 2016; Glass and Cook 2018; Hoyt and Murphy 2016; Kaiser and Wallace 2016; Simmons and Yawson 2022). For example, the continuation of existing disparities in career advancement and equal opportunities for WOC not only hinders their personal and professional growth but also poses a significant setback to the overall diversity, equity, and inclusion objectives of organizations (Acker 2006; Barak 2016; Bawany 2022; Bloch, Taylor, Church, and Buck 2020; Carli 2020; Diehl and Dzubinski 2016; Fitzsimmons and Callan 2016; Glass and Cook 2018; Hoyt and Murphy 2016; Kaiser and Wallace 2016; Muñoz 2020; Simmons and Yawson 2022).

Furthermore, neglecting the specific challenges faced by WOC in the workplace can result in missed opportunities for innovation, collaboration, and productivity (Acker 2006; Barak 2016; Bawany 2022; Bloch, Taylor, Church, and Buck 2020; Carli 2020; Diehl and Dzubinski 2016; Fitzsimmons and Callan 2016; Glass and Cook 2018; Hoyt and Murphy 2016; Kaiser and Wallace 2016; Muñoz 2020; Simmons and Yawson 2022). As a result, the failure to address these issues could lead to a continued underrepresentation of diverse perspectives at leadership levels, limiting the holistic success and adaptability of organizations, especially during times of disruptive change (Becker and Aiello 2013; Carter and Sisco 2022; Eagly and Heilman 2016; Everly and Athey 2023; Hewlett 2008; Kim 2020; Moody, Lewis, and Owens 2022; Phipps and Prieto 2020; Potter 2021; Rincón Diez González, and Barrero 2017; Spates, Evans, James, and Martinez 2020). Therefore, investigating and implementing effective strategies for the career development of WOC is not only a matter of social justice but also a crucial factor for fostering a resilient and thriving workforce in the face of ongoing challenges (Ellemers 2014; Lee and Wessel 2022; Moody, Lewis, and Owens 2022; Snaebjornsson, Edvardsson, Zydziunaite, and Vaiman 2015).

Gendered Racism in Corporate America

Advocating for a seat at the table and climbing hills to attain equity and equality in the workplace are metaphorical depictions of the career advancement obstacles faced by women, especially WOC, in corporate America because of *gendered racism* (Becker and Aiello 2013; Carter and Sisco 2022; Eagly and Heilman 2016; Kim 2020; Moody, Lewis, and Owens 2022; Phipps and Prieto 2020; Rincón Diez González, and Barrero 2017; Spates, Evans, James, and Martinez 2020; Williams and Wiggins 2010). Race and gender issues in America are an ongoing societal debate about whether equity and equality in the workplace should be set as business priorities or left to individual responsibility (Ellemers 2014; Moody, Lewis, and Owens 2022; Pilcher and Whelehan 2004; Smith 2000; Snaebjornsson, Edvardsson, Zydziunaite, and Vaiman 2015). However, despite the personal and career challenges WOC encounter in reaching self-actualization of career achievement, they still have the perseverance and capability to succeed (Bowen 2021; Eagly and Heilman 2016; Eisele and Stake 2008; Maslow 2000; Rincón Diez González, and Barrero 2017).

Career Advancement Challenges

Through qualitative studies, researchers have found that WOC in the United States do not reach career advancement at the same pace as men or White women, including equitable pay or access to leadership development opportunities (Moody, Lewis, and Owens 2022; Oleschuk 2020; Snaebjornsson, Edvardsson, Zydziunaite, and Vaiman 2015). There is also limited research on how cross-cultural perspectives, including race, age, socioeconomic status, and sexual orientation, impact WOC in the workplace (Bonacini, Gallo, and Scicchitano 2021; Carter and Sisco 2022; Crichlow, Coburn, and Joseph 2020; Shim, Dik, and Banning 2022; Smith 2000). For example, most research on women in the workforce focuses on studying *all* women as one group without considering WOC as a targeted population with different needs and life experiences (Bonacini, Gallo, and Scicchitano 2021; Carter and Sisco 2022; Crichlow, Coburn, and Joseph 2020). Thus, a knowledge gap precludes the diverse perspectives of WOC in the workplace, which deviates from the progress made on race and gender advocacy over the last 100 years due to the negative impacts of disruptive societal changes, such as the 2020 COVID-19 pandemic, economic recessions, and gendered racism (Acker 2006; Bawany 2022; Becker and Aiello 2013; Crichlow, Coburn, and Joseph 2020; Cubrich et al. 2022; Kim 2020; Kim and Gege 2021; Landivar, Ruppanner, Scarborough, and Collins 2020; McClain 2020; Moody, Lewis, and Owens 2022; Mukkamala and Suyemoto 2018; O'Conner 1996; Oleschuk 2020; Phipps and Prieto 2020; Pilcher and Whelehan 2004; Randsley de Moura et al. 2018; Smith 2000; Snyder 2008; Spates, Evans, James, and Martinez 2020).

Moreover, WOC will continue to face socioeconomic barriers, including homeownership and financial security, thereby precluding them from attaining higher education goals, leadership roles, and peer equity in the workplace (Acker 2006; Bem 1994; Carter and Sisco 2022; Eagly and Heilman 2016; Ellemers 2014; Ibarra, Ely, and Kolb 2019; Oleschuk 2020; Smith 2000; Ziglar 2021). As a result, the conflicts created by the lack of race and gender diversity in corporate America, especially in organizational leadership roles, will continue to negatively impact proven positive societal, organizational, and individual contributions of WOC to the workforce, specifically during times of disruptive societal change (Acker 2006; Barak 2016; Bawany 2022; Bloch, Taylor, Church, and Buck 2020; Boulouta 2012; Carli 2020; Diehl and Dzubinski 2016; Fitzsimmons and Callan 2016; Glass and Cook 2018; Hoyt and Murphy 2016; Kaiser and Wallace 2016; Muñoz 2020; Pilcher and Whelehan 2004; Simmons and Yawson 2022; Ziglar 2021).

The Implications of Disruptive Change

Studying disruptive change and the intersectionality of work–life integration, gender and race relations, socioeconomics, and career development resources for WOC are instrumental to studying this problem of practice (Bawany 2022; Denson, Szelényi, and Bresonis 2018; Diehl and Dzubinski 2016; Everly and Athey 2023; Fitzsimmons and Callan 2016; Fritz and van Knippenberg 2018; Hoyt and Murphy 2016; Kaiser and Wallace 2016; Kalysh, Kulik, and Perera 2016; Lee and Wessel 2022; Paustian-Underdahl et al. 2017; Pilcher and Whelehan 2004; Potter 2021; Warner and Shields 2018; Wayne, Matthews, Crawford, and Casper 2020). Therefore, more analysis is needed to discern how organizational leaders can integrate cross-culturally focused strategies to support career advancement in the workplace for WOC, especially during disruptive changes (Acker 2006; Bawany 2022; Bloch, Taylor, Church, and Buck 2020; Carli 2020; Diehl and Dzubinski 2016; Ellemers 2014; Everly and Athey 2023; Fitzsimmons and Callan 2016; Glass and Cook 2018; Hoyt and Murphy 2016; Kaiser and Wallace 2016; Moody, Lewis, and Owens 2022; Potter 2021; Simmons and Yawson 2022; Smith 2000).

Conceptual Framework

The conceptual framework guiding this study was constructed leveraging efficacy theory, the empowerment model of practice, feminist theory, and Maslow's hierarchy of needs. WOC encounter a spectrum of challenges, spanning interpersonal, financial, social, and societal domains, particularly within their professional contexts (Acker 2006; Dietz 2003; Durand et al. 2021; Feng and Savani 2020; Fitzsimmons and Callan 2016; Jackson and Jones 1998; Lyness and Grotto 2018; McClain 2020; Robnett and Anderson 2017; Wayne, Matthews, Crawford, and Casper 2020). WOC have encountered diverse challenges, spanning interpersonal, financial, social, and societal domains, specifically in their professional domains (Acker 2006; Dietz 2003; Durand et al. 2021; Feng and Savani 2020; Fitzsimmons and Callan 2016; Jackson and Jones 1998; Lyness and Grotto 2018; McClain 2020; Robnett and Anderson 2017; Wayne, Matthews, Crawford, and Casper 2020). Therefore, each concept contributed to the framework, offering a comprehensive understanding of how various elements intersected with race and gender inequity for WOC in the United States. The subsequent sections provide a detailed exploration of each theory, contributing to a comprehensive understanding of how these diverse elements intersected with race and gender inequity for WOC in the United States.

Efficacy Theory

Efficacy theory states that powerlessness and oppressive structures impact one's sense of power and capabilities in the world around them (Bandura 1982; Gutiérrez and Lewis 1999; Mahalingam 2007). According to this theory, powerless people experience decreased self-efficacy because of negative societal experiences and interactions (Bandura 1984; Gutiérrez and Lewis 1999; Mahalingam 2007). However, some people view powerlessness as a challenge that requires more grit and resiliency to overcome obstacles (Gutiérrez and Lewis 1999; Mahalingam 2007; Turner,

Ekachai, and Slattery 2022). Therefore, one's beliefs about power structures and social behaviors can change through positive experiences that increase influential capabilities and feelings of self-efficacy (Bandura 1984; Gutiérrez and Lewis 1999; Mahalingam 2007). Under this premise, WOC would feel more competent in gaining power from historically White male-dominated corporate power structures (Gutiérrez and Lewis 1999; Humm 1995; Kolmar and Bartkowski 2000; Mahalingam 2007; Northouse 2018). As a result, WOC who feel confident, supported, and included in organizational teams and programs would likely experience increased positive workplace experiences (Chen 2014; Gutiérrez and Lewis 1999; Humm 1995; Kolmar and Bartkowski 2000; Northouse 2018).

Empowerment Model of Practice

The empowerment model of practice focuses on how various societal groups gain and leverage power and control over distributed resources (Gutiérrez and Lewis 1999). The empowerment model typically focuses on increasing social justice while eliminating inequities that reduce power for historically marginalized groups (Atkins and Duckworth 2019; Aydarova 2019; Durand et al. 2021; Gutiérrez and Lewis 1999; Mahalingam 2007). Thus, WOC empowered by their organizations to progress in their careers can also gain more power over their financial health and well-being (Chen et al. 2014; Durand et al. 2021; Gutiérrez and Lewis 1999; Mahalingam 2007; Taormina and Gao 2013; Vermote et al. 021).

Feminist Theory

Feminist theory provides a narrative explanation and analysis of women's lives (Dietz 2003; Jackson and Jones 1998; Kolmar and Bartkowski 2000). A core precept of feminist theory is that power structures, such as male-dominated leadership groups, negatively impact women based on gender, race, class, sexuality, nationality, and other implied or stated differences (Dietz 2003; Jackson and Jones 1998; Kolmar and Bartkowski 2000; Mahalingam 2007; Pilcher and Whelehan 2004).

Maslow's hierarchy of needs Maslow's hierarchy of needs supports the conceptual framework by outlining how people are motivated by basic needs using a hierarchical model represented by a pyramid (Maslow 1943). The hierarchy model consists of a progressive scale of needs inclusive of the following:

- physiological: food, water, shelter
- psychological: safety
- emotional: love, belonging
- esteem: confidence, worthiness
- self-actualization: purpose

As a result, a person will likely not reach the top of the hierarchy or self-actualization if they are hungry or feel unsafe (Maslow 1943). To attain fulfillment, people must procure basic physiological needs, such as security, safety, social affiliation, esteem, and self-actualization (Goler, Gale, Harrington, and Grant 2018; Maslow 1943). Therefore, each tier in the hierarchy is relative to one another but distinct in what must occur to move up from the bottom or first physiological tier (Maslow 1943). Furthermore, the data collected and analyzed for the study could inform researchers about how disruptive societal changes in corporate America negatively or positively impact WOC on each level of Maslow's hierarchy.

The History of Women of Color and Gendered Racism in the United States

Historically, when disruptive changes infiltrated society in the United States, women were subjected to increased levels of socially constructed gender norms that affected their career advancement opportunities and financial stability (Bawany 2022; Bruhn 2022; Carli 2020; Chavez-Dueñas, Adames, and Perez-Chavez 2022; Collins, Landivar, Ruppanner, and Scarborough 2020; Dias, Chance, and Buchanan 2020; Everly and Athey 2023; Kim and Gege 2021; Landivar, Ruppanner, Scarborough, and Collins 2020; Pilcher and Whelehan 2004; Potter 2021; Power 2020; Shockley, Clark, Dodd, and King 2021). More specifically, WOC bore the significant and often negative impacts of disruptive societal changes (Acker 2006; Bawany 2022; Bloch, Taylor, Church, and Buck 2020; Bruhn 2022; Gutiérrez and Lewis 1999; Mukkamala and Suyemoto 2018; Paustian-Underdahl et al. 2017).

For example, WOC experienced more challenges integrating family and career responsibilities, which contributed to substantial race and gender inequalities in the workplace during the recent 2020 COVID-19 pandemic, which was a substantial disruptive societal event (Bawany 2022; Cubrich et al. 2022; Lyttelton, Zang, and Musick 2021; Pilcher and Whelehan 2004; Whiley, Sayer, and Juanchich 2021). Additionally, men and women have been conditioned to adhere to predisposed gender norms in socially normalized power structures due to dominant theoretical perspectives on the social construction of gender (Becker and Aiello 2013; Bem 1994; Kim 2020; Kolmar and Bartkowski 2000; Mahalingam 2007; Moody, Lewis, and Owens 2022; Pilcher and Whelehan 2004; Snyder 2008; Spates, Evans, James, and Martinez 2020). As a result, women must cope with gender-based stereotypes in their workplaces (Bem 1994; Kolmar and Bartkowski 2000; Moody, Lewis, and Owens 2022; Snyder 2008).

Feminist theory identifies and examines conceptual topics, such as gender and socioeconomic inequality (Bem 1994; Dietz 2003; Jackson and Jones 1998; Oleschuk 2020; Pilcher and Whelehan 2004; Ridgeway 2014). For example, in the early days of the women's suffrage movement, the feminist majority did not view WOC as activists or scholarly contributors (McCammon, Campbell, Granberg, and Mowery 2001; Nicholson 1997; Organization of American Historians 2019). Moreover, research on women's suffrage has highlighted how racism became entwined in the movement led by White middle-class women as strategic bait to gain the right to vote (McCammon, Campbell, Granberg, and Mowery 2001; Nicholson 1997; Organization of American Historians 2019).

Some feminist theorists described how early women's suffrage activists focused solely on White middle-class women and less on working-class women and WOC who were less likely to vote (McCammon, Campbell, Granberg, and Mowery 2001; Nicholson 1997; Organization of American Historians 2019). As a result, WOC leveraged churches and plantations to lay the groundwork for political suffrage separate from White women activists (McCammon, Campbell, Granberg, and Mowery 2001; Nicholson 1997; Organization of American Historians 2019; Robnett and Anderson 2017). Thus, narrowing down themes in social situations and narratives that triggered early gender and race issues for WOC in America benefited the study (Bem 1994; McCammon, Campbell, Granberg, and Mowery 2001; Oleschuk 2020; Pilcher and Whelehan 2004).

Moreover, women, especially WOC, took on most, if not all, of the childcare and household duties versus men, even in dual-income households where both caregivers were telecommuters (Alon et al. 2021; Carli 2020; Collins, Landivar, Ruppanner, and Scarborough 2020; Dias, Chance, and Buchanan 2020; Hays 1996; Kim and Gege 2021; Landivar, Ruppanner, Scarborough, and Collins 2020; Power 2020; Shockley, Clark, Dodd, and King 2021). Furthermore, current research indicates that single mothers of color are more socially isolated than married mothers of color, work longer hours, receive less emotional support, and have less stable social networks (Hays 1996; Taylor et al. 2022). As a result, single mothers of color face intensified adversities, which unfavorably impact their mental health and parenting skills (Hays 1996; Taylor et al. 2022). Thus, a deeper review of how WOC face challenges through disruptive societal changes from various cultures and historically disenfranchised backgrounds requires further investigation.

The Road to Women's Suffrage

Before American colonialization and the formation of the United States, Indigenous women (IW) in the Americas were honored and valued for their positive contributions to culture and society (Chavez-Dueñas, Adames, and Perez-Chavez 2022; Coburn 2020). For example, IW had property rights and financial security regardless of marriage (Chavez-Dueñas, Adames, and Perez-Chavez 2022; Coburn 2020;). However, once colonialism began, the ideology of patriarchy took over, where men were superior to women and all Indigenous cultures (Chavez-Dueñas, Adames, and Perez-Chavez 2022; Coburn 2020). Additionally, colonialism ignited the atrocities of slavery in the Americas, which negatively impacted an entire culture of African people for centuries to come, especially African American women (Carter and Rossi 2019; Coburn 2020; Moody, Lewis, and Owens 2022; Nasar 2020; Spates, Evans, James, and Martinez 2020).

Enslaved African American or Black women experienced ongoing, sordid oppression ranging from physical, verbal, and emotional abuse to sexual assault and exploitation (Carter and Rossi 2019; Coburn 2020; Collins 1992; Moody, Lewis, and Owens 2022; Nasar 2020; Spates, Evans, James, and Martinez 2020). Moreover, Black women had to care for their enslaver's families, which included an array of domestic duties and chores, in addition to caring for other enslaved people, enslaved children, and other compulsory labor (Carter and Rossi 2019; Coburn 2020; Collins 1992; Moody, Lewis, and Owens 2022; Nasar 2020; Spates, Evans, James, and Martinez 2020). As a result, the stereotype of the strong Black woman emerged despite the consistent marginalization and dehumanization these women faced daily (Carter and Rossi 2019; Coburn 2020; Moody, Lewis, and Owens 2022; Nasar 2020; Spates, Evans, James, and Martinez 2020; Waldron 2019). Current research indicates that Black women experience gendered racism at higher rates than other WOC, which also increases the likelihood of health issues, including chronic stress, hypertension, diabetes, mental illness, and other physical and emotional ailments (Becker and Aiello 2013; Carter and Rossi 2019; Collins 1992; Moody, Lewis, and Owens 2022; Mukkamala and Suyemoto 2018; Smith 2000; Waldron 2019; Williams and Wiggins 2010).

Women of Mexican descent, either born in Mexico or the United States, are historically overrepresented in low paying jobs where career mobility is limited or nonexistent, such as clerical work, housekeeping, childcare, and food services (Abji, Korteweg, and Williams 2019; Coburn 2020; Garcini, Rodríguez, Mercado, and Paris 2020; Taylor et al. 2022). Due to cultural norms and stereotypes, familial responsibilities continue to impact their ability to secure higher paying jobs or advance their education (Coburn 2020; Taylor et al. 2022). Additionally, Mexican women who are undocumented immigrants must navigate harsher disruptive changes that complicate their lives due to highly volatile and politically charged immigration laws and exploitation (Abji, Korteweg, and Williams 2019; Bawany 2022; Bruhn 2022; Cárdenas 1975; Coburn 2020; Everly and Athey 2023; Garcini, Rodríguez, Mercado, and Paris 2020; Jones 2021; Massey 2020; Moya 2015; Potter 2021). Furthermore, Indigenous women experience gendered racism in the United States and get typecast into one group, such as Latina or Native American, although there is an array of Indigenous cultures globally (Becker and Aiello 2013; Chavez-Dueñas, Adames, and Perez-Chavez 2022; Coburn 2020; DeBlaere et al. 2023; Guzmán and Valdivia 2004; Mukkamala and Suyemoto 2018; Taylor et al. 2022).

Similarly, Asian American women experience gendered racism due to the disparities of traditional Asian stereotypes in Western culture (Becker and Aiello 2013; Kim 2020; Lee et al. 2022; Li 2013; Mukkamala and Suyemoto 2018). For example, Asian American women take care of family and household duties while also working outside the home to provide financial support (Alon et al. 2021; Carli 2020; Collins, Landivar, Ruppanner, and Scarborough 2020; Dias, Chance, and Buchanan 2020; Kim and Gege 2021; Landivar, Ruppanner, Scarborough, and Collins 2020; Lee et al. 2022; Li 2013; Power 2020; Shockley, Clark, Dodd, and King 2021).

Moreover, the Asian culture includes diverse ethnic groups, including Chinese, Filipino, Indian, Japanese, Korean, Taiwanese, Vietnamese, and other Asian countries (Alon et al. 2021; Carli 2020; Collins, Landivar, Ruppanner, and Scarborough 2020; Dias, Chance, and Buchanan 2020; Kim and Gege 2021; Landivar, Ruppanner, Scarborough, and Collins 2020; Lee et al. 2022; Li 2013; Mukkamala and Suyemoto 2018; Power 2020; Shockley,

Clark, Dodd, and King 2021). Thus, many first-generation Asian American women struggle to identify with Western values and manage the expectations of their cultural heritage, which varies by country (Alon et al. 2021; Carli 2020; Collins, Landivar, Ruppanner, and Scarborough 2020; Dias, Chance, and Buchanan 2020; Kim and Gege 2021; Landivar, Ruppanner, Scarborough, and Collins 2020; Lee et al. 2022; Li 2013; Mukkamala and Suyemoto 2018; Power 2020; Shockley, Clark, Dodd, and King 2021).

A stereotype of Asian women might be that they are submissive or only work at nail salons (Alon et al. 2021; Carli 2020; Collins, Landivar, Ruppanner, and Scarborough 2020; Dias, Chance, and Buchanan 2020; Kim and Gege 2021; Landivar, Ruppanner, Scarborough, and Collins 2020; Lee et al. 2022; Li 2013; Mukkamala and Suyemoto 2018; Power 2020; Shockley, Clark, Dodd, and King 2021). As a result, all Asian women become categorized as one group versus subgroups with diverse cultural norms, which leads to distorted stereotypes that negatively impact their career advancement (Alon et al. 2021; Carli 2020; Collins, Landivar, Ruppanner, and Scarborough 2020; Dias, Chance, and Buchanan 2020; Kim and Gege 2021; Landivar, Ruppanner, Scarborough, and Collins 2020; Lee et al. 2022; Li 2013; Mukkamala and Suyemoto 2018; Power 2020; Shockley, Clark, Dodd, and King 2021).

The Women's Suffrage Movement

Research and historical analysis of the women's suffrage movement described how societal norms forced women into domestic roles while men dominated business and political arenas (Hymowitz and Weissman 1978; McClain 2020; O'Conner 1996). However, pioneers of the suffrage movement in the United States, such as Susan B. Anthony and Maggie Lena Walker, challenged discrepancies in women's rights, such as voting privileges (Case, Iuzzini, and Hopkins 2012; McClain 2020; Phipps and Prieto 2020). For example, women were considered inferior to men by nature, probably less intelligent, and biologically unsuited to the pressures of work and politics (Hymowitz and Weissman 1978; McClain 2020; O'Conner 1996; Rojas 2009). Furthermore, feminist scholars demonstrated how gender, racial, and class inequality patterns differ depending on the type of economic activity present in various American regions (Acker 2006; Hymowitz and Weissman 1978; Pilcher and Whelehan 2004; Robnett and Anderson 2017; McClain 2020; Rojas 2009). Thus, understanding what types of disruptive societal changes have occurred in corporate America since the enactment of the 19th Amendment in 1920 provides historical information on the issues WOC continue to face in corporate America (Acker 2006; Bawany 2022; Hymowitz and Weissman 1978; O'Conner 1996; Rojas 2009).

For example, WOC, specifically Black women, had to fight even harder for gender equity versus White women due to the ongoing racial inequality of Black Americans (Collins 1992; Conrad, Dixson, Smooth, and Revilla 2014; Johnson 2021; Khanmalek 2020; McClain 2020; Rojas 2009; Smith 2000). Sojourner Truth, a notable figure of Black women's rights, led the charge with her notorious speech, *Ain't I A Woman?*, which set the tone for how Black women struggled with gender equality within an already marginalized group (Collins 1992; Hymowitz and Weissman 1978; McClain 2020; Pilcher and Whelehan 2004; Smith 2000). Moreover, feminist scholars of color have contended for decades that feminist literature focuses more on White middle-class women and less on the complicated history of gender, race, and socioeconomic status for WOC (Acker 2006; Hymowitz and Weissman 1978; Khanmalek 2020; Pilcher and Whelehan 2004; Robnett and Anderson 2017; Rojas 2009; Shim, Dik, and Banning 2022). As a result, gender and race equity for WOC in corporate America continues to be a long-term battle for socioeconomic and career advancement (Johnson 2021; Khanmalek 2020; McClain 2020; Pilcher and Whelehan 2004; Redfern 1989; Rojas 2009).

The Transformation of Feminism in the United States

Feminist researchers have explored why women continue to be underrepresented in leadership roles or placed in high-risk positions where their credibility is compromised (Randsley de Moura et al. 2018; Redfern 1989). Many feminist theorists believe there are transformational periods of feminist thought that emerged in the United States after women won the right to vote (Falconi et al. 2020; Friedan 2001; Friedman 2005; Goldin 1991; Hall, Orzada, and Lopez-Gydosh 2015; Nicholson 1997; Organization of American Historians 2019; Robnett and Anderson 2017). The first period occurred from the 1920s to the early 1960s when women pushed for rights to enter the paid workforce and forgo solely domestic and nonpaid roles, also known as the women's movement (Goldin 1991; Nicholson 1997; Falconi et al. 2020; Friedman 2005; Hall, Orzada, and Lopez-Gydosh 2015; Organization of American Historians 2019).

During World War II, many women left their domestic roles to join the workforce out of necessity and patriotism, causing the most significant influx of US female labor in modern history (Goldin 1991; Friedman 2005; Hall, Orzada, and Lopez-Gydosh 2015; Falconi et al. 2020). Female labor increased by over 35% from 1940 to 1945 during WW II, while men served in the armed forces (Falconi et al. 2020; Friedman 2005; Goldin 1991; Hall, Orzada, and Lopez-Gydosh 2015; Nicholson 1997; Organization of American Historians 2019). Thus, a societal shift in traditional norms occurred once WW II started, placing women as head of household, replacing historically patriarchal systems in the United States (Falconi et al. 2020; Friedman 2005; Goldin 1991; Hall, Orzada, and Lopez-Gydosh 2015).

Additionally, WOC had higher employment rates than White women, but only in professions with lower educational requirements (Falconi et al. 2020; Friedman 2005; Goldin 1991; Hall, Orzada, and Lopez-Gydosh 2015;). Regardless of race, women now had two jobs—domestic duties and paid wage labor (Falconi et al. 2020; Friedan 2001; Friedman 2005; Goldin 1991; Hall, Orzada, and Lopez-Gydosh 2015). However, after WW II ended, gender role conflicts ignited as laws and regulations caused women to return home to solely domestic duties (Falconi et al. 2020; Friedan 2001; Friedman 2005; Goldin 1991; Hall, Orzada, and Lopez-Gydosh 2015; Pilcher and Whelehan 2004). Many critics contended that allowing women to remain in the workforce threatened job availability for men (Falconi et al. 2020; Friedan 2001; Friedman 2005; Goldin 1991; Hall, Orzada, and Lopez-Gydosh 2015).

For example, middle-class and married women were publicly ostracized for continuing to work, with some states banning employers from hiring married women or barring them from paid work altogether (Falconi et al. 2020; Friedan 2001; Friedman 2005; Goldin 1991; Hall, Orzada, and Lopez-Gydosh 2015). In contrast, WOC continued to work outside the home to survive financially, often working multiple blue-collar jobs to support their families (Falconi et al. 2020; Friedan 2001; Friedman 2005; Goldin 1991; Hall, Orzada, and Lopez-Gydosh 2015; Meisenbach 2009). As a result, many proponents of the women's movement contended that neither women's suffrage nor WW II addressed issues of race and class for WOC in the workplace (Organization of American Historians 2019).

The next movement followed in the late 1960s when the notion of oppressive power structures emerged as a focus for women's liberation (Mahalingam 2007; Nicholson 1997). For example, feminists argued that women were entitled to satisfy material needs similar to men "in the context of social structures that changed over time" (Nicholson 1997: 2). Consequently, society's view of men and women as equitable became a highly contested debate among feminists and other opposition groups (Nicholson 1997; Robnett and Anderson 2017; Taramundi 2016). For example, sex discrimination cases increased in the US courts under the premise that men and women were entitled to the same financial benefits related to pay, property, and childcare obligations (Nicholson 1997). However, other sex-based legislation upheld the male-only military draft and statutory rape laws (Nicholson 1997). As a result, many societal gender norms continued to impact case law in proceeding feminist movements, which fueled the debate on gender equity and equality in the United States (Conrad, Dixson, Smooth, and Revilla 2014; Nicholson 1997; Pilcher and Whelehan 2004; Ridgeway 2014; Taramundi 2016; Robnett and Anderson 2017).

Modern-age feminist movements addressed generational conflicts from the prior feminist periods, highlighting how male-dominated power structures still existed to the detriment of gender and race activism (Andersen and Collins 2004; Leicht et al. 2017; Mahalingam 2007; Pilcher and Whelehan 2004; Snyder 2008). For example, male social power in capitalist societies, such as the United States, stems from socially accepted norms of competition, rationalization, and domination in corporate settings (Andersen and Collins 2004; Leicht et al. 2017; Mahalingam 2007; Nicholson 1997). Thus, women who oppose gender norms and behave as men in corporate settings are deemed irrational and emotional (Andersen and Collins 2004; Leicht et al. 2017; Nicholson 1997; Pilcher and Whelehan 2004). As a result, men continue to prevail as the ideal corporate figures (Andersen and Collins 2004; Leicht et al. 2017; Nicholson 1997). At the same time, research showed that women fell behind in career advancement and leadership development based on socially accepted gender norms in a male-dominated corporate America (Andersen and Collins 2004; Leicht et al. 2017; Nicholson 1997). Therefore, dominant societal gender norms continue to prevail in corporate America to the detriment of modern feminist agendas (Andersen and Collins 2004; Leicht et al. 2017; Nicholson 1997).

Gender and Racial Power in US Society

Gender and race are often predictors of power in US society due to White male-dominated representation in political and business arenas (Andersen and Collins 2004; Donovan 1992; Hoyt and Murphy 2016; Mahalingam 2007; Pilcher and Whelehan 2004; Shockley, Clark, Dodd, and King 2021). Historically, White men possessed power over women across all economic and ethnic lines (Bruhn 2022; Mahalingam 2007; McClain 2020; Rojas 2009; SHRM 2021; Women in the Workplace 2023). As a result, socially constructed cultural values and gender roles gave more power to men over women and people of color and denied them equal rights (Andersen and Collins 2004; Donovan 1992; Hoyt and Murphy 2016; Mahalingam 2007; Rojas 2009; Snaebjornsson, Edvardsson, Zydziunaite, and Vaiman 2015). For example, men justified prejudices against women and people of color with no reason or cause during the late 1700s (Wollstonecraft 2019). Additionally, women were codependent on men and did not have power over their bodies or lives (Donovan 1992; Mahalingam 2007; McClain 2020). As a result, women labeled *feminists* were deemed less credible and subjected to social stigmatization by men and women who subscribed to normalized societal gender stereotypes (Brescoll 2016; Kleinman 2007; Pilcher and Whelehan 2004; Rojas 2009; SHRM 2021; White 2006; Women in the Workplace 2023).

At the same time, WOC remained misrepresented in feminist literature because female stereotypes were inconsistent with their life experiences (Eagleton 1986; Rojas 2009). Thus, the 1920s women's suffrage movement did little to eradicate race discrimination for WOC (Donovan 1992; McClain 2020; Rojas 2009). As a result, WOC experienced more challenges in gaining equitable access to the same rights as White women (Donovan 1992; Eagleton 1986; Rojas 2009; White 2006;). Moreover, proponents of feminist advocacy implored all women to learn, understand, and share their history or risk losing hard-fought freedoms, which laid the groundwork for modern feminist movements (Donovan 1992; Eagleton 1986; Kleinman 2007; Rojas 2009; White 2006).

Gender and Race in Leadership

The *think manager–think male paradigm* describes how society perceives women as lacking emotional intelligence to make successful managers or leaders (Fischbach, Lichtenthaler, and Horstmann 2015; Halliday, Paustian-Underdahl, and Fainshmidt 2020). Moreover, women leaders are viewed differently depending on established societal and corporate cultural norms (Fischbach, Lichtenthaler, and Horstmann 2015; Halliday, Paustian-Underdahl, and Fainshmidt 2020; Hewlett 2008). For example, women encounter more penalties in the workplace by displaying behaviors that emulate nontraditional gender norms, such as asking for a promotion or pay increase (Denson, Szelényi, and Bresonis 2018; Eagly and Heilman 2016; Fritz and van Knippenberg 2018; Women in the Workplace 2023). As a result, women leaders are perceived differently based on established societal gender norms (Denson, Szelényi, and Bresonis 2018; Eagly and Heilman 2016; Fritz and van Knippenberg 2018; Women in the Workplace 2023). Perceptions of leadership legitimacy from historically White male-dominated business domains are often significantly lower for women, negatively impacting their career progression (Halliday, Paustian-Underdahl, and

Fainshmidt 2020; Vial, Napier, and Brescoll 2016; Women in the Workplace 2023). As a result, women leaders are at a perpetually biased disadvantage due to socially dominated views of gender behaviors (Lyness and Grotto 2018; Pilcher and Whelehan 2004; Randsley de Moura et al. 2018; Ryan et al. 2016; Vial, Napier, and Brescoll 2016; Women in the Workplace 2023).

However, many US organizations have attempted to decrease race and gender gaps in recruitment and leadership development through DEI initiatives (Barak 2016; Lyness and Grotto 2018; Muñoz 2020; Randsley de Moura et al.; Ryan et al. 2016). Nonetheless, women leaders continue to experience more prevalent discrimination than their male counterparts regarding compensation, performance evaluation, and career development opportunities (Lee and Wessel 2022; Lyness and Grotto 2018; Randsley de Moura et al. 2018; Ryan et al. 2016). Thus, organizations need to understand how to implement effective DEIB programs and resources that target the underlying causes of inequitable access to career advancement opportunities for WOC, specifically in higher-level leadership roles (Barak 2016; Carter and Sisco 2022; Chen 2014; Harvard Business Review Analytic Services 2021; Women in the Workplace 2023).

For example, research on US business cultures revealed how women were more likely to be promoted into leadership positions during times of crisis or instability (Carter and Sisco 2022; Women in the Workplace 2023). Moreover, social norms regarding gender stereotypes contributed to women appearing incapable or unsuited for leadership (Brescoll 2016; Fischbach, Lichtenthaler, and Horstmann 2015; Halliday, Paustian-Underdahl, and Fainshmidt 2020; Pilcher and Whelehan 2004; Randsley de Moura et al. 2018; Rosette, Koval, Ma, and Livingston 2016). However, researchers have sought to expand the current work surrounding gender and race diversity in the workplace (Barak 2016; Eagly and Heilman 2016; Muñoz 2020). Additionally, research into race and gender in US leadership showed how interventions to increase women's representation in leadership roles would be more effective if guided by social research and peer-reviewed data (Eagly and Heilman 2016; Gurung 2020; Randsley de Moura et al. 2018; Women in the Workplace 2023). Conversely, research on women in leadership has focused primarily on White women and fails to consider intersectional identities and cross-cultural challenges for WOC (Brandert and Matkin 2019; Carter and Sisco 2022; Eagly and Carli 2019; Rosette, Koval, Ma, and Livingston 2016; Ryan et al. 2016; Women in the Workplace 2023). As a result, WOC are categorized with all women in scholarly research, even though WOC have significantly different life experiences and access to resources in corporate America (Carter and Sisco 2022; Rosette, Koval, Ma, and Livingston 2016; Ryan et al. 2016; Women in the Workplace 2023).

Modern advocates for gender equity in the workplace, such as former Facebook COO Sheryl Sandberg, are strong proponents of *all* women getting a seat at the table alongside a predominantly White male workforce (Conrad, Dixson, Smooth, and Revilla 2014; Phipps and Prieto 2020; Pilcher and Whelehan 2004; Women in the Workplace 2023). For example, women often score higher on leadership effectiveness models yet remain in inferior positions to men (Fitzsimmons and Callan 2016; Kaiser and Wallace 2016; Women in the Workplace 2023). Moreover, negative stereotypes of women devalue and negate the positive attributes and contributions that female diversity brings to leadership teams (Barak 2016; Ellemers 2014; Ellemers, Rink, Derks, and Ryan 2012; Paustian-Underdahl et al. 2017). As a result, women with behavioral traits of competitiveness or aggressiveness are penalized for not conforming to traditional gender norms (Ellemers, Rink, Derks, and Ryan 2012; Pilcher and Whelehan 2004; Women in the Workplace 2023). In contrast, women who exemplify more social-emotional leadership behaviors, such as benevolence and empathy, are viewed as incompetent leaders compared to their male peers who show traditional male-dominated, competitive gender behaviors (Ellemers 2014; Ellemers, Rink, Derks, and Ryan 2012; Pilcher and Whelehan 2004; Women in the Workplace 2023). Thus, gender-based stereotypes create challenges for women, especially WOC, and career development opportunities (Ellemers 2014; Ellemers, Rink, Derks, and Ryan 2012; Lee and Wessel 2022; Paustian-Underdahl et al. 2017; Pilcher and Whelehan 2004; Women in the Workplace 2023).

Equity Versus Inequality in the Modern Workplace

Women continue to get paid less than men, who control most human capital strategies, even though the pay gap between the genders has narrowed (Albelda and Tilly 1997; Eagly and Heilman 2016; Pilcher and Whelehan 2004; Power 2020; Women in the Workplace 2023). Thus, women have lower wages, status positions, overscheduling, limited paid time off, and higher rates of mistreatment by supervisors (Albelda and Tilly 1997; Bruhn 2022; Eagly and Heilman 2016; Power 2020; Women in the Workplace 2023). WOC also experience unequal access to power and resources in society because of racism, sexism, and classism in America (Albelda and Tilly 1997; Gutiérrez and Lewis 1999; Mahalingam 2007; SHRM 2021; Women in the Workplace 2023). As a result, modern disruptive changes in the workplace, such as the 2020 COVID-19 pandemic, were conduits for gender and pay inequities for WOC (Bawany 2022; Christensen, McDonald, Altman, and Palmer 2018; Cubrich et al. 2022; Everly and Athey 2023; Ozkazanc-Pan and Pullen 2020; Pilcher and Whelehan 2004; Potter 2021; SHRM 2021; Simmons and Yawson 2022; Volini et al. 2019; Women in the Workplace 2023; Yawson 2020).

Additionally, social injustice events in the United States caused organizations to redefine the ideal business culture and focus more resources on fostering culturally diverse workplaces (Oleschuk 2020; SHRM 2021; Women in the Workplace 2023). However, even when organizational policies and programs positively impacted and improved the lives of all women, WOC were often left behind, as is the case that prompted the Women's Education Act of 1974 (Chen 2014; Conrad, Dixson, Smooth, and Revilla 2014). Thus, WOC remain the most impacted during disruptive societal changes in the US workplace (Acker 2006; Bawany 2022; Bloch, Taylor, Church, and Buck 2020; Bruhn 2022; Paustian-Underdahl et al. 2023).

Recent research has also indicated that WOC were more likely to be single mothers with limited financial support and resources (Albelda and Tilly 1997; Bruhn 2022; Taylor et al. 2022). For example, WOC experienced poverty at more than 50% the rate of White women in the United States during the 2020 COVID-19 pandemic (Marsh-McDonald and Schroeder 2012; Ozkazanc-Pan and Pullen 2020; US Census Bureau 2021). Furthermore, WOC continue to face systemic gender and race inequality due to negative stereotypes in the workplace that hinder career progression (Bonacini, Gallo, and Scicchitano 2021; Kalysh, Kulik, and Perera 2016; Pilcher and Whelehan 2004; Ridgeway 2014; SHRM 2021; Women in the Workplace 2023). As a result, disruptive societal changes that organizational leaders do not effectively manage increase the socioeconomic gap for WOC, especially in the workplace (Bawany 2022; Bloch, Taylor, Church, and Buck 2020; Bonacini, Gallo, and Scicchitano 2021; Bruhn 2022; Crichlow, Coburn, and Joseph 2020; SHRM 2021; Women in the Workplace 2023).

Disruptive Changes in the Business World

Disruptive change is a part of business and society that accelerates globalization, communications, and technological innovation (Ahn, Adamson, and Dornbusch 2004; Bawany 2022; Everly and Athey 2023; Potter 2021). The changing patterns of cross-border capital flows and the fluid state of corporate mergers and partnerships create fundamental shifts in how business is conducted in the United States (Ahn, Adamson, and Dornbusch 2004). Consequently, disruptive changes also introduce more adversity in the workplace that negatively impacts the advancement of gender and race equity (Bawany 2022; Bloch, Taylor, Church, and Buck 2020; Bruhn 2022; Eagly and Heilman 2016; Everly and Athey 2023; Pilcher and Whelehan 2004; Potter 2021; SHRM 2021). For example, organizational leaders are typically unprepared when disruptions impact the workplace, especially for WOC (Ahn, Adamson, and Dornbusch 2004; Bawany 2022; Ganesh and Zoller 2012; SHRM 2021). However, organizations that develop leaders who improvise and embrace inclusivity provide more options for managing disruptive workplace changes that affect employees (Bawany 2022; SHRM 2021; Simmons and Yawson 2022; Women in the Workplace 2023).

Disruptions Due to the 2020 COVID-19 Pandemic

The 2020 COVID-19 pandemic was a *macrodisruption* to societies around the world, defined as a sizeable disruptive event followed by multiple smaller microdisruptions, such as a lack of childcare and remote working (Bawany 2022; Bruhn 2022; Cubrich et al. 2022; Kerr, Rasmussen, Fanning, and Braaten 2021; Ozkazanc-Pan and Pullen 2020; Turner, Ekachai, and Slattery 2022; Zhang et al. 2020). During the 2020 COVID-19 pandemic, a recent and significant disruptive change, a large population of women, especially WOC, left their careers to care for children, elderly parents, or for their own health concerns (Bawany 2022; Carli 2020; Cubrich et al. 2022; Everly and Athey 2023; Ozkazanc-Pan and Pullen 2020; Potter 2021; Simmons and Yawson 2022). The mass exodus of working women, coined the *2020 Shecession*, created a problem for the modern workplace due to the decreased female and racial representation in leadership and other key roles, which limited innovation and cross-cultural collaboration (Bonacini, Gallo, and Scicchitano 2021; Crichlow, Coburn, and Joseph 2020; Cubrich et al. 2022; Hegewisch and Mefferd 2021; Paweenawat and Liao 2021; SHRM 2021; Soto 2020; Ziglar 2021). As a result, organizational stakeholders and leaders faced challenges in understanding how to make gender and race equity and inclusivity a business priority during and in the aftermath of the 2020 COVID-19 pandemic (Cubrich et al. 2022; Ozkazanc-Pan and Pullen 2020; Pitt-Catsouphes et al. 2022; SHRM 2021; Simmons and Yawson 2022).

Microdisruptions reported from a study of 24 working American mothers exemplified how finances, job duties, and the lack of time and space to work added stress to an already traumatic situation (Turner, Ekachai, and Slattery 2022). Moreover, WOC were impacted at higher rates due to strained financial resources, lack of childcare, and other socioeconomic components (Pandey 2021; SHRM 2021). For example, WOC incurred more "systemic barriers, particularly those around family and household responsibilities" (Ibarra, Ely, and Kolb 2019; Oleschuk 2020: 503). Women, especially WOC, assumed more than double the household and childcare duties of men, creating higher stress levels for working mothers (Feng and Savani 2020; Fitzsimmons and Callan 2016; Forbes, Lamar, and Bornstein 2020; Lyness and Grotto 2018; Lyttelton, Zang, and Musick 2021; Wayne, Matthews, Crawford, and Casper 2020). Furthermore, WOC faced increased career disadvantages versus their peers, which reduced the likelihood of attaining senior leadership and C-suite positions, notwithstanding the regular quest for career progression and advancement opportunities (Feng and Savani 2020; Fitzsimmons and Callan 2016; Lyness and Grotto 2018; Moe and Shandy 2010; Oleschuk 2020; SHRM 2021; Wayne, Matthews, Crawford, and Casper 2020). Thus, WOC faced more challenges managing their career and family obligations than their non-WOC peers (Moe and Shandy 2010; Oleschuk 2020; Wayne, Matthews, Crawford, and Casper 2020).

However, choosing career progression over caring for a family has been a long-standing issue for most women since before the 2020 COVID-19 pandemic (Carli 2020; Cubrich et al. 2022; Ladge, Humberd, and Eddleston 2018; Moe and Shandy 2010; Ozkazanc-Pan and Pullen 2020; Wayne, Matthews, Crawford, and Casper 2020). For example, women, specifically WOC, tend to have less independent financial security, educational attainment, and career advancement opportunities versus men and non-WOC due to disruptive societal changes (Bawany 2022; Feng and Savani 2020; Fitzsimmons and Callan 2016; Lyness and Grotto 2018; SHRM 2021; Wayne, Matthews, Crawford, and Casper 2020). Additionally, many WOC left the job market, which created a problem for diversity, equity, and inclusion efforts due to decreased female and racial representation (Barak 2016; Muñoz 2020; Pitt-Catsouphes et al. 2022; SHRM 2021; Simmons and Yawson 2022). As a result, organizational stakeholders and leaders faced challenges in understanding how to make gender and race inclusivity a business priority in a post-pandemic business landscape (SHRM 2021; Simmons and Yawson 2022). However, research showed that organizational stakeholders and leaders consistently challenged the investment requested by people leaders, such as HR professionals, of making gender and race inclusivity a business priority until after the 2020 COVID-19 pandemic began (Ozkazanc-Pan and Pullen 2020; SHRM 2021; Simmons and Yawson 2022).

Furthermore, the adverse effects of the 2020 COVID-19 pandemic exacerbated lower-income families, including a lack of healthcare, paid leave, and opportunities to work from home, mainly for WOC (Cubrich et al. 2022; Kerr,

Rasmussen, Fanning, and Braaten 2021; Ozkazanc-Pan and Pullen 2020; SHRM 2021). For example, the 2020 COVID-19 pandemic's effects led to almost a third of US women receiving either a pay cut or furlough, which doubled for WOC (Cubrich et al. 2022; Ozkazanc-Pan and Pullen 2020; Power 2020; SHRM 2021). As a result, WOC lost career development opportunities, which increased race and gender inequality in corporate America (Bonacini, Gallo, and Scicchitano 2021; Crichlow, Coburn, and Joseph 2020; Cubrich et al. 2022; Diehl and Dzubinski 2016; Denson, Szelényi, and Bresonis 2018; Kim and Gege 2021; Landivar, Ruppanner, Scarborough, and Collins 2020; Lee and Wessel 2022; Oleschuk 2020; SHRM 2021). Thus, women of all cross-cultural backgrounds were tasked with more competing priorities at higher rates than men, contributing to increased gender disparity in career advancement opportunities (Hoyt and Murphy 2016; Oleschuk 2020; SHRM 2021).

According to the US Bureau of Labor Statistics, over 5.8 million jobs were vacated by women, accounting for 53.9% of overall job loss since the start of the 2020 COVID-19 pandemic (Crichlow, Coburn, and Joseph 2020; Möhring, Reifenscheid, Weiland, and Kuhn 2021; Ozkazanc-Pan and Pullen 2020; US Bureau of Labor Statistics 2021). Additionally, 1.5 million people lost their jobs within the first two months of the pandemic, leading to the lowest employment rate for women in 30 years (Crichlow, Coburn, and Joseph 2020; Cubrich et al. 2022). In contrast, about 500,000 new jobs were added to the US workforce in May of 2021, with 56% belonging to women (Hegewisch and Mefferd 2021). However, employment rates for WOC remained below pre-COVID levels and did not accurately represent the percentage of employed women (Cubrich et al. 2022; Hegewisch and Mefferd 2021). Additional research indicated that 42% of women worked part-time during the pandemic (Crichlow, Coburn, and Joseph 2020; Cubrich et al. 2022; Möhring, Reifenscheid, Weiland, and Kuhn 2021). In comparison, only 6.4% of men worked part-time, emphasizing how household duties or service sector cuts impacted women, especially WOC, at higher rates (Crichlow, Coburn, and Joseph 2020; Cubrich et al. 2022; Möhring, Reifenscheid, Weiland, and Kuhn 2021).

During the 2020 COVID-19 pandemic, 9.89% of women were unemployed, which increased for WOC (Cubrich et al. 2022; Crichlow, Coburn, and Joseph 2020; Hegewisch and Mefferd 2021; Möhring, Reifenscheid, Weiland, and Kuhn 2021; Ozkazanc-Pan and Pullen 2020; Simmons and Yawson 2022). For example, Black women had a 1.7% higher unemployment rate than White women. Hispanic/Latina women experienced a 1.5% higher unemployment rate (Cubrich et al. 2022; Goode et al. 2022; Guzmán and Valdivia 2004; Hegewisch and Mefferd 2021.). Moreover, the 2020 COVID-19 uncovered how Black women had a distinctly different experience due to racial capitalism and misogyny, which is defined as a deep-rooted prejudice against Black women (Collins 1992; Crichlow, Coburn, and Joseph 2020; Cubrich et al. 2022; Goode et al. 2022; Johnson 2021; SHRM 2021; Smith 2000). Therefore, a job recession is more likely to impact female-dominated sectors due to the tendency of WOC to work for lower pay (Collins 1992; Crichlow, Coburn, and Joseph 2020; Cubrich et al. 2022; Goode et al. 2022; Möhring, Reifenscheid, Weiland, and Kuhn 2021; SHRM 2021; Simmons and Yawson 2022).

Furthermore, unemployment statistics from 2020 and 2021 showed how the United States experienced a significant increase in unemployment for women due to a lack of contingency planning regarding disruptive changes for diverse populations in corporate America (Bawany 2022; Crichlow, Coburn, and Joseph 2020; Cubrich et al. 2022; Everly and Athey 2023; Möhring, Reifenscheid, Weiland, and Kuhn 2021; Potter 2021; SHRM 2021; Simmons and Yawson 2022). As a result, the 2020 COVID-19 magnified preexisting vulnerabilities for women, especially WOC, of low to middle socioeconomic statuses while concurrently adding new stressors created by the pandemic (Crichlow, Coburn, and Joseph 2020; Cubrich et al. 2022; D'Auria and De Smet 2020; Goode et al. 2022; Kim and Gege 2021; Möhring, Reifenscheid, Weiland, and Kuhn 2021; SHRM 2021; Shim, Dik, and Banning 2022).

On another note, the unemployment crisis caused by the 2020 COVID-19 pandemic increased household duties and unwanted pregnancies for women, especially WOC (Crichlow, Coburn, and Joseph 2020; Cubrich et al. 2022; D'Auria and De Smet 2020; Goode et al. 2022; Kim and Gege 2021; Möhring, Reifenscheid, Weiland, and Kuhn

2021). For example, household duties increased by 10% for women and was higher for WOC, with nearly 30% of pregnant women quitting their jobs due to the perceived higher risk of catching a viral infection (Bonacini, Gallo, and Scicchitano 2021; Crichlow, Coburn, and Joseph 2020; D'Auria and De Smet 2020; Kim and Gege 2021; Möhring, Reifenscheid, Weiland, and Kuhn 2021). Additionally, Black and minority ethnic women experienced increased challenges balancing career and childcare obligations, putting them at a higher risk for health complications from COVID-19 and other ailments due to increased stress and physical strain (D'Auria and De Smet 2020; Goode et al. 2022; SHRM 2021; Whiley, Sayer, and Juanchich 202). Furthermore, approximately 60% of women worked in noncorporate sectors, such as retail and hospitality, where they were more likely to become impoverished during the pandemic, with more impact on WOC (D'Auria and De Smet 2020; Goode et al. 2022; Kim and Gege 2021; SHRM 2021). As a result, WOC experienced additional stresses and challenges in navigating career, family, and mental health issues during the 2020 COVID-19 pandemic (Cubrich et a. 2022; D'Auria and De Smet 2020; Goode et al. 2022; Hegewisch and Mefferd 2021; Kalysh, Kulik, and Perera 2016; Ladge, Humberd, and Eddleston 2018; Moe and Shandy 2010; SHRM 2021; Wayne, Matthews, Crawford, and Casper 2020).

Work–Life Integration and Organizational Resources

Historically, women have been stigmatized for choosing career advancement over family responsibilities (Hegewisch and Mefferd 2021; Kalysh, Kulik, and Perera 2016; Ladge, Humberd, and Eddleston 2018; Moe and Shandy 2010; O'Conner 1996; SHRM 2021; Wayne, Matthews, Crawford, and Casper 2020; Zhang et al. 2020). Moreover, the demands of managing work duties and familial obligations prevented women from pursuing or being considered for career development opportunities (Bear and Glick 2017; Diehl and Dzubinski 2016; Lee and Wessel 2022; Moe and Shandy 2010; SHRM 2021; Zhang et al. 2020). Thus, in traditional gender role societies like the United States, women are expected to care for the home and children versus pursuing a career (Moe and Shandy 2010; SHRM 2021; Whiley, Sayer, and Juanchich 2021; Zanhour and Sumpter 2022). As a result, women experience more pressure to choose a family over a career during periods of disruptive change (Bawany 2022; Everly and Athey 2023; Potter 2021; SHRM 2021; Whiley, Sayer, and Juanchich 2021; Zanhour and Sumpter 2022; Zhang et al. 2020).

In April 2020, the United Nations reported a significant increase in the unpaid care economy stemming from a lack of childcare resources and increased household duties due to school and daycare closures caused by public health mandates (Alon et al. 2021; Collins, Landivar, Ruppanner, and Scarborough 2020; Dias, Chance, and Buchanan 2020; Fisher and Ryan 2021; Landivar, Ruppanner, Scarborough, and Collins 2020; Oleschuk 2020; Power 2020; SHRM 2021). As a result, the boundaries between work and home impacted the ability for WOC to maintain sufficient employment and financial stability (Fisher and Ryan 2021; Lyttelton, Zang, and Musick 2021; SHRM 2021; Whiley, Sayer, and Juanchich 2021; Zanhour and Sumpter 2022). Disruptive societal changes further heightened the stress and family responsibilities for working mothers of color due to financial insecurity, layoffs, and limited childcare resources (Alon et al. 2021; Bawany 2022; Bloch, Taylor, Church, and Buck 2020; Bruhn 2022; Carli 2020; Fisher and Ryan 2021; Forbes, Lamar, and Bornstein 2020; Hegewisch and Mefferd 2021; Oleschuk 2020; Pandey 2021; Power 2020).

In contrast, White women with more available childcare and financial resources experienced higher work–life satisfaction than WOC (Oleschuk 2020; Wayne, Matthews, Crawford, and Casper 2020; SHRM 2021; Westring and Ryan 2011). As a result, working mothers of color were more challenged in choosing between their careers and familial priorities by employers who failed to provide work–life integration support and resources (Fisher and Ryan 2021; Forbes, Lamar, and Bornstein 2020; Lyttelton, Zang, and Musick 2021; Meisenbach 2009; Oleschuk 2020; Wayne, Matthews, Crawford, and Casper 2020; Westring and Ryan 2011; Whiley, Sayer, and Juanchich 2021; Zanhour and Sumpter 2022.). Furthermore, the concept of an ideal worker is based on male-centric views that work comes first (SHRM 2021; Whiley, Sayer, and Juanchich 2021; Zanhour and Sumpter 2022). Thus, the ideal worker mindset creates additional barriers for working mothers, especially WOC (Fisher and Ryan 2021; Forbes, Lamar, and

Bornstein 2020; Ibarra, Ely, and Kolb 2019; Lyttelton, Zang, and Musick 2021; Meisenbach 2009; SHRM 2021; Whiley, Sayer, and Juanchich 2021; Zanhour and Sumpter 2022.).

In multiple studies, researchers explored how implementing work–life integration initiatives directly correlated to an increased number of women in leadership roles (Brandert and Matkin 2019; Eagly and Carli 2019; Fritz and van Knippenberg 2018; SHRM 2021; Westring and Ryan 2011). Moreover, companies that provided various options and resources to promote work–life integration, such as flex schedules and on-site childcare, retained working mothers at higher rates (Forbes, Lamar, and Bornstein 2020; Fritz and van Knippenberg 2018; Meisenbach 2009; SHRM 2021). Thus, working mothers were able to increase their bandwidth, allowing them to continue career advancement and manage family obligations, resulting in more gender diversity in corporate America (Barak 2016; Boulouta 2012; Forbes, Lamar, and Bornstein 2020; Kim and Gege 2021; Ladge, Humberd, and Eddleston 2018; Pandey 2021; SHRM 2021; Whiley, Sayer, and Juanchich 2021). As a result, the stigma and negative stereotypes of women-dominated childcare reduced gender inequality in those workplaces (Fritz and van Knippenberg 2018; Forbes, Lamar, and Bornstein 2020; Ladge, Humberd, and Eddleston 2018; SHRM 2021; Wayne, Matthews, Crawford, and Casper 2020).

Mental Health and Well-Being

Maslow's hierarchy of needs provides relevant context to the study of WOC in the workplace by explaining how disruptive societal changes impact a person's hierarchical needs (Bawany 2022; Goler, Gale, Harrington, and Grant 2018; Hill-Jarrett and Jones 2021; Maslow 1943, 1998, 2000, 2013; SHRM 2021). If WOC encounter barriers to achieving lower levels of need, such as food, water, shelter, and financial security, then the upper levels of self-esteem, peer recognition, and self-actualization are not met (Fedi and Rollero 2016; Goler, Gale, Harrington, and Grant 2018; Ibarra, Ely, and Kolb 2019; Maslow 1943, 1998, 2000, 2013). For example, numerous psychology-based studies determined that the need for belonging affects people in every culture and social space, including family, friendships, and work relationships (Chen et al. 2014; Goler, Gale, Harrington, and Grant 2018; Hill-Jarrett and Jones 2021; Maslow 1998, 2000, 2013; Taormina and Gao 2013; SHRM 2021).

A modern view of Maslow's hierarchical approach to human needs indicates that workplaces fulfill not only basic needs, but all the needs listed in the pyramid (Bowen 2021; Goler, Gale, Harrington, and Grant 2018; Maslow 1998, 2000, 2013). For example, an analysis of organizations that employees noted as the best places to work from 1984 to 2011 showed increased profitability of 2.3% to 3.8% yearly (Bowen 2021; Goler, Gale, Harrington, and Grant 2018; Maslow 1998, 2000, 2013). The research highlighted how companies that invested in satisfying their employee's needs gained more revenue than their competitors (Bowen 2021; Goler, Gale, Harrington, and Grant 2018; Maslow 1998, 2000, 2013). Moreover, three main themes stood out from the research, indicating how people are motivated by career, community, and cause, notated as unwritten expectations in the form of a *psychological contract* (Bowen 2021; Goler, Gale, Harrington, and Grant 2018; Maslow 1998, 2000, 2013). As a result, employers that comply with the psychological contract experience higher employee engagement and productivity (Bowen 2021; Goler, Gale, Harrington, and Grant 2018; Maslow 1998, 2000, 2013). Specific psychological needs, such as relatedness, competence, and autonomy, impact people in various ways (Bowen 2021; Chen et al. 2014; Goler, Gale, Harrington, and Grant 2018; Hill-Jarrett and Jones 2021; Maslow 1998, 2000, 2013; Taormina and Gao 2013). Therefore, self-esteem, popularity influence, and safety are required to support one's psychological needs (Bowen 2021; Chen et al. 2014; Fedi and Rollero 2016; Goler, Gale, Harrington, and Grant 2018; Hill-Jarrett and Jones 2021; Maslow 1998, 2000, 2013; SHRM 2021; Taormina and Gao 2013). Thus, without the necessary psychological support, people become defensive, insecure, physically ill, and psychopathic (Bowen 2021; Chen et al. 2014; Goler, Gale, Harrington, and Grant 2018; Hill-Jarrett and Jones 2021; Maslow 1998, 2000, 2013; SHRM 2021; Taormina and Gao 2013).

During a disruptive societal change, such as the 2020 COVID-19 pandemic, a lack of psychological support increased stress, worry, and insecurity for WOC because basic needs were unmet (Bawany 2022; Cubrich et al. 2022;

D'Auria and De Smet 2020; Goler, Gale, Harrington, and Grant 2018; Hill-Jarrett and Jones 2021; Lewis 2023; Moody, Lewis, and Owens 2022; Vance, Wade, Brandy, and Webster 2022; Vermote et al. 2021). For example, lockdown measures isolated WOC, leaving them powerless to meet their basic needs due to decreased human contact, resources, and support (Goler, Gale, Harrington, and Grant 2018; Hill-Jarrett and Jones 2021; Lewis 2023; Mahalingam 2007; Moody, Lewis, and Owens 2022; Seubert, Hopfgartner, and Glaser 2021; Vermote et al. 2021). Moreover, the fear of contracting COVID-19 was disproportionately higher for WOC, which increased feelings of isolation and stress (Carli 2020; D'Auria and De Smet 2020; Goler, Gale, Harrington, and Grant 2018; Hill-Jarrett and Jones 2021; Lewis 2023; Moody, Lewis, and Owens 2022; SHRM 2021; Vance, Wade, Brandy, and Webster 2022). Additionally, most WOC worked essential jobs requiring more contact with the public, including healthcare, cleaning services, and food preparation (Carli 2020; D'Auria and De Smet 2020; Goler, Gale, Harrington, and Grant 2018; Hill-Jarrett and Jones 2021; Lewis 2023; Moody, Lewis, and Owens 2022; Vance, Wade, Brandy, and Webster 2022). Thus, the predominance of WOC serving as essential workers placed them at a greater risk of illness, stress, burnout, and impaired mental health (Carli 2020; D'Auria and De Smet 2020; Goler, Gale, Harrington, and Grant 2018; Hill-Jarrett and Jones 2021; Lewis 2023; Moody, Lewis, and Owens 2022; SHRM 2021; Vance, Wade, Brandy, and Webster 2022).

Recent studies have indicated that adults with minimal social support experience higher rates of depression than adults with more support (Carter and Rossi 2019; Hill-Jarrett and Jones 2021; Moody, Lewis, and Owens 2022; Racine et al. 2021; Vance, Wade, Brandy, and Webster 2022). For example, WOC with lower incomes and who are primarily single and uneducated mothers experienced lower levels of gratitude and self-esteem, which decreased mental health and well-being (Carter and Rossi 2019; Fedi and Rollero 2016; Goler, Gale, Harrington, and Grant 2018; Hill-Jarrett and Jones 2021; Lewis 2023; Moody, Lewis, and Owens 2022; Racine et al. 2021; Taylor et al. 2022). Furthermore, mothers dealing with continuous stress experienced more burnout, negatively impacting parenting behaviors and increasing exhaustion and emotional distress (Carter and Rossi 2019; Goler, Gale, Harrington, and Grant 2018; Hill-Jarrett and Jones 2021; Kerr, Rasmussen, Fanning, and Braaten 2021; Lewis 2023; Moody, Lewis, and Owens 2022; Szymanski and Stewart 2010; Vance, Wade, Brandy, and Webster 2022). Consequently, mothers of young children felt higher levels of physical and emotional distress due to the loss of social support and childcare resources, which led to increased economic hardship (Carter and Rossi 2019; Goler, Gale, Harrington, and Grant 2018; Hill-Jarrett and Jones 2021; Lewis 2023; Moody, Lewis, and Owens 2022; Racine et al. 2021; Szymanski and Stewart 2010; Vance, Wade, Brandy, and Webster 2022). In a study of 1,009 parents from the beginning of the 2020 COVID-19 pandemic, 39.4% reported experiencing higher anxiety and depression, with at least one symptom of burnout reported by 69% of participating parents (Cubrich et al. 2022; Kerr, Rasmussen, Fanning, and Braaten 2021). Thus, the negative impact on basic physiological needs for WOC prevented the satisfaction of higher-level psychological needs and placed the potential for self-actualization at a disadvantage (Carter and Rossi 2019; Goler, Gale, Harrington, and Grant 2018; Hill-Jarrett and Jones 2021; Lewis 2023; Moody, Lewis, and Owens 2022; Racine et al. 2021; Taormina and Gao 2013).

Alternate Viewpoints

The literature on gender equity and inequality has raised questions about the United States' progress in closing the gender gap (Conrad, Dixson, Smooth, and Revilla 2014; Lyness and Grotto 2018; Pitt-Catsouphes et al. 2022; SHRM 2021; Women in the Workplace 2023). For example, scholarly research has focused on the challenges preventing women from advancing into leadership ranks (Lyness and Grotto 2018; SHRM 2021; Women in the Workplace 2023). However, during a study of gender bias and leadership ratings, researchers found minimal evidence of bias in favor of women leaders, which was insufficient to substantiate a strong correlation (Kaiser and Wallace 2016; SHRM 2021; Women in the Workplace 2023). Additionally, scholarly research remains deficient in examining cross-cultural disparities for women in leadership, specifically for underrepresented populations, such as the LBGTQ+

community and WOC (Brandert and Matkin 2019; Eagly and Carli 2019; Hoyt and Murphy 2016; McClain 2020; SHRM 2021; Women in the Workplace 2023). Thus, while current research exists on women in leadership, the studies focused mainly on gender and less on racial equity and equality in the workplace for WOC (Brandert and Matkin 2019; Eagly and Carli 2019; Hoyt and Murphy 2016; McClain 2020; Pitt-Catsouphes et al. 2022; SHRM 2021; Women in the Workplace 2023).

Conclusion

The focus of this research was that WOC in America consistently bear the brunt of personal and professional repercussions during disruptive societal crises such as pandemics, race relations, political unrest, and economic recessions (Bawany 2022; Carli 2020; Collins, Landivar, Ruppanner, and Scarborough 2020; Cubrich et al. 2022Dias, Chance, and Buchanan 2020; Fisher and Ryan 2021). These disruptive consequences compound race, gender, and socioeconomic inequities for WOC, leading to obstacles in their career progression (Bawany 2022; Carli 2020; Collins, Landivar, Ruppanner, and Scarborough 2020; Dias, Chance, and Buchanan 2020; Fisher and Ryan 2021; Pilcher and Whelehan 2004). Furthermore, the adverse outcomes experienced by WOC in US workplaces due to gendered racism are a culmination of shifting workforce demands, strained resources, and societal instability (Alon et al. 2021; Becker and Aiello 2013; Bonacini, Gallo, and Scicchitano 2021; Carli 2020; Collins, Landivar, Ruppanner, and Scarborough 2020; Hegewisch and Mefferd 2021; Hoyt and Murphy 2016; Kim 2020; Mukkamala and Suyemoto 2018; Pandey 2021; Simmons and Yawson 2022; Spates, Evans, James, and Martinez 2020).

However, during crises, such as sociopolitical unrest, more WOC find themselves in high profile leadership roles where failure is more likely (Carter and Sisco 2022; Dias, Chance, and Buchanan 2020; Diehl and Dzubinski 2016; Fitzsimmons and Callan 2016; Hoyt and Murphy 2016; Lyness and Grotto 2018; McClain 2020; Paustian-Underdahl et al. 2017; Randsley de Moura et al. 2018). Moreover, WOC must navigate the career implications of gendered racism, bias, isolation, tokenism, microaggressions, and financial disparity in the workplace (Acker 2006; Becker and Aiello 2013; Bloch, Taylor, Church, and Buck 2020; Carter and Sisco 2022; DeBlaere et al. 2023; Kim 2020; Lewis and Neville 2015; Lyness and Grotto 2018; McClain 2020; Mukkamala and Suyemoto 2018; Randsley de Moura et al. 2018; Spates, Evans, James, and Martinez 2020; Sue et al. 2007).

Adding to the urgency for this problem to be studied is that research on WOC in leadership is notably scarce (Bloch, Taylor, Church, and Buck 2020; Carter and Sisco 2022; Diehl and Dzubinski 2016; Fitzsimmons and Callan 2016; Hoyt and Murphy 2016; Lyness and Grotto 2018; McClain 2020; Randsley de Moura et al. 2018). Most studies examining women in leadership broadly do not explicitly address the nuances in cross-cultural dynamics for WOC as a subset (Brandert and Matkin 2019; Bloch, Taylor, Church, and Buck 2020; Carter and Sisco 2022; Diehl and Dzubinski 2016; Eagly and Carli 2019; Fitzsimmons and Callan 2016; Hoyt and Murphy 2016; Kaiser and Wallace 2016; Lyness and Grotto 2018; McClain 2020; Randsley de Moura et al. 2018). Consequently, there is a significant gap in studies on organizational strategies facilitating the progression of WOC into various careers and leadership roles (Acker 2006; Bloch, Taylor, Church, and Buck 2020; Carter and Sisco 2022; Diehl and Dzubinski 2016; Fitzsimmons and Callan 2016; Hoyt and Murphy 2016; Kaiser and Wallace 2016; Lyness and Grotto 2018; McClain 2020; Randsley de Moura et al. 2018). As a result, this lack of insight perpetuates the existing disparities for WOC, thus limiting the potential positive contributions of WOC in the workplace.

Thus, this research aimed to investigate the career journeys faced by WOC amid disruptive changes, exploring both professional and personal dimensions. The goal was to advocate for practices that facilitate more significant career advancement and leadership opportunities, thereby supporting the personal and professional growth of WOC. The findings from this study have the potential to inform and shape strategies that promote inclusivity and equality in the business landscape (Barak 2016; Cinamon 2006; Durand et al. 2021; Ellemers 2014; Gray and O'Brien 2007; Hideg and Shen 2019; Lowman 2005; Swanson 2022).

References

Abji, Salina, Anna C. Korteweg, and Lawrence E. Williams. 2019. "Culture talk and the politics of the new right: Navigating gendered racism in attempts to address violence against women in immigrant communities." *Signs, Vol. 44*, no. 3, pp. 797-822. https://doi.org/10.1086/701161

Acker, Joan. (2006). "Inequality regimes." *Gender and Society, Vol.* 20, no. 4, pp. 441-64. https://doi.org/10.1177/0891243206289499

Ahn, Mark J., John S.A. Adamson, and Daniel Dornbusch. 2004. "From leaders to leadership: Managing change." *Journal of Leadership and Organizational Studies, Vol. 10*, no. 4, pp. 112-23. https://doi.org/10.1177/107179190401000409

Albelda, Randy, and Chris Tilly. 1997. *Glass Ceilings and Bottomless Pits: Women's Work, Women's Poverty*. South End Press.

Alon, Titan, Sena Coskun, Matthias Doepke, Davud Koll, and Michele Tertilt. 2021. "From mancession to shecession: Women's employment in regular and pandemic recessions." *NBER Macroeconomics Annual, Vol. 36*, no. 1, 83-151. https://doi.org/10.1086/718660

Andersen, Margaret L., and Patricia H. Collins. (2004). *Race, Class, and Gender: An Anthology* (5th ed.). Wadsworth Publishing Company.

Atkins, Liz, and Vicky Duckworth. 2019. *Research Methods for Social Justice and Equity in Education*. Springer. https://doi.org/10.1007/978-3-030-05900-2

Aydarova, Elena. 2019. Flipping the paradigm: Studying up and research for social justice. In K. Strunk and L. A. Locke (Eds.), *Research Methods for Social Justice and Equity in Education* (pp. 33–43). Palgrave Macmillan. https://doi.org/10.1007/978-3-030-05900-2_3

Bandura, Albert. 1982. "The assessment and predictive generality of self-percepts of efficacy." *Journal of Behavior Therapy and Experimental Psychiatry, Vol. 13*, no. 3, 195-99. https://doi.org/10.1016/0005-7916(82)90004-0

Bandura, Albert. 1984. "Recycling misconceptions of perceived self-efficacy." *Cognitive Therapy and Research, Vol. 8*, no. 3, pp. 231-55. https://doi.org/10.1007/bf01172995

Barak, Michalle E. Mor. 2016. *Managing Diversity: Toward a Globally Inclusive Workplace*. Sage.

Bawany, Sattar. 2022. *Leadership in Disruptive Times: Negotiating the New Balance* (2nd ed.). Business Expert Press.

Bear, Julia B., and Peter Glick. 2017. "Breadwinner bonus and caregiver penalty in workplace rewards for men and women." *Social Psychological and Personality Science, Vol. 8*, no. 7, pp. 780-88. https://doi.org/10.1177/1948550616683016

Becker, Sarah, and Brittnie Aiello. 2013. "The continuum of complicity: "Studying up"/studying power as a feminist, anti-racist, or social justice venture." *Women's Studies International Forum, Vol. 38*, pp. 63-74. https://doi.org/10.1016/j.wsif.2013.02.004

Bem, Sandra L. 1994. *The Lenses of Gender: Transforming the Debate on Sexual Inequality*. Yale University Press.

Bloch, Katrina R., Tiffany Taylor, Jacob Church, and Alison Buck. 2020. "An intersectional approach to the glass ceiling: Gender, race and share of middle and senior management in U.S. workplaces." *Sex Roles, Vol. 84*, no. 5-6, pp. 312-25. https://doi.org/10.1007/s11199-020-01168-4

Bonacini, Luca, Giovanni Gallo, and Sergio Scicchitano. 2021. "Will it be a she-cession? The unintended influence of working from home on the gender wage gap related to the COVID-19 pandemic." *GLO Discussion Paper, Vol. 771*, pp. 1-31. https://www.econstor.eu/handle/10419/229149

Boulouta, Ioanna. 2012. "Hidden connections: The link between board gender diversity and corporate social performance." *Journal of Business Ethics, Vol. 113*, no. 2, pp. 185-97. https://doi.org/10.1007/s10551-012-1293-7

Bowen, Bob. 2021. "The matrix of needs: Reframing Maslow's hierarchy." *Health, Vol. 13*, no. 05, pp. 538-63. https://doi.org/10.4236/health.2021.135041

Brandert, Kathleen T., and Gina S. Matkin. 2019. "When the crisis is personal: A phenomenological study of women in leadership." *Journal of Leadership Studies, Vol. 13*, no. 3, pp. 56-61. https://doi.org/10.1002/jls.21663

Brescoll, Victoria L. 2016. "Leading with their hearts? How gender stereotypes of emotion lead to biased evaluations of female leaders." *Leadership Quarterly, Vol. 27*, no. 3, pp. 415-28. https://doi.org/10.1016/j.leaqua.2016.02.005

Bruhn, Sarah. 2022. "Me cuesta mucho": Latina immigrant mothers navigating remote learning and caregiving during COVID-19." *Journal of Social Issues, Vol. 79*, no. 3, 1-22. https://doi.org/10.1111/josi.12546

Cárdenas, Gilberto. 1975. "United States immigration policy toward Mexico: A historical perspective." *Chicana/o-Latina/O Law Review, Vol. 2*, 66-92. https://doi.org/10.5070/c720020914

Carli, Linda L. 2020. "Women, gender equality and COVID-19." *Gender in Management: An International Journal, 35*(7-8), 647–655. https://doi.org/10.1108/gm-07-2020-0236

Carter, Angela D., and Stephanie Sisco. 2022. "The hill we climb: The future of women of color leadership development." *Advances in Developing Human Resources, Vol. 24*, no. 3, pp. 208-19. https://doi.org/10.1177/15234223221099665

Carter, Leeja, and Amerigo Rossi. 2019. "Embodying strength: The origin, representations, and socialization of the strong Black woman ideal and its effect on Black women's mental health." *Women and Therapy, Vol. 42*, no. 3-4, pp. 289-300. https://doi.org/10.1080/02703149.2019.1622911

Case, Kim A., Jonathan Iuzzini, and Morgan Hopkins. 2012. "Systems of privilege: Intersections, awareness, and applications." *Journal of Social Issues, Vol. 68*, no. 1, 1-10. https://doi.org/10.1111/j.1540-4560.2011.01732.x

Chavez-Dueñas, Nayeli Y., Hector Y. Adames, and Jessica G. Perez-Chavez. 2022. "Anti-colonial futures: Indigenous Latinx women healing from the wounds of racial-gendered colonialism." *Women and Therapy, Vol. 45*, no. 2-3, pp. 191-206. https://doi.org/10.1080/02703149.2022.2097593

Chen, Beiwen, Jasper Van Assche, Maarten Vansteenkiste, Bart Soenens, and Wim Beyers. 2014. "Does psychological need satisfaction matter when environmental or financial safety are at risk?" *Journal of Happiness Studies, Vol. 16*, 745-66. https://doi.org/10.1007/s10902-014-9532-5

Chen, Huey T. 2014. *Practical Program Evaluation: Theory-Driven Evaluation and the Integrated Evaluation Perspective* (2nd ed.). Sage.

Christensen, Clayton M., Rory McDonald, Elizabeth J. Altman, and Jonathan E. Palmer. 2018. "Disruptive innovation: An intellectual history and directions for future research." *Journal of Management Studies, Vol. 55*, no. 7, pp. 1043-78. https://doi.org/10.1111/joms.12349

Cinamon, Rachel G. 2006. "Anticipated work-family conflict: Effects of gender, self-efficacy, and family background." *Career Development Quarterly, Vol. 54*, no. 3, pp. 202-15. https://doi.org/10.1002/j.2161-0045.2006.tb00152.x

Coburn, Elaine. 2020. "Theorizing our place": Indigenous women's scholarship from 1985-2020 and the emerging dialogue with anti-racist feminisms." *Studies in Social Justice, Vol. 14*, no. 2, pp. 429-53. https://doi.org/10.26522/ssj.v14i2.2295

Collins, Caitlyn, Liana Christin Landivar, Leah Ruppanner, and William J. Scarborough. 2020. "COVID-19 and the gender gap in work hours." *Gender, Work and Organization, Vol. 28*, no. S1, pp. 101-12. https://doi.org/10.1111/gwao.12506

Collins, Patricia H. 1992. "Black feminist thought: Knowledge, consciousness, and the politics of empowerment." *Contemporary Sociology, Vol. 21*, no. 1, p. 132. https://doi.org/10.2307/2074808

Conrad, Cecilia, Adrienne Dixson, Wendy Smooth, and Anita Tijerina Revilla. 2014. "A discussion on gender equity and women of color." *Frontiers: A Journal of Women Studies, Vol. 35*, no. 3, pp. 3-14. https://doi.org/10.1353/fro.2014.0037

Crichlow, Wesley, Elaine Coburn, and Emma Joseph. 2020. "The "she-cession" and Black womanhood during the pandemic: What is to be done?" *Antipode: A Radical Journal of Geography*, pp. 1–6. https://antipodeonline.org/wp-content/uploads/2020/12/4.-Crichlow-Coburn-and-JosLeeweph.pdf

Cubrich, Marc, Joren A. Tengesdal, Gustavo Ugueto-Rey, Rachael Stahl, and Megan Crow Brauer. 2022. "Pandemics and precarious work: Translating research to practice for marginalized workers." *Translational Issues in Psychological Science, Vol. 8*, no. 3, pp. 416-30. https://doi.org/10.1037/tps0000327

D'Auria, Gemma, and Aaron De Smet. 2020. *Leadership in a Crisis: Responding to the Coronavirus Outbreak and Future Challenges*. McKinsey and Company. https://www.mckinsey.com/business-functions/organization/our-insights/leadership-in-a-crisis-responding-to-the-coronavirus-outbreak-and-future-challenges#

DeBlaere, Cirleen, David G. Zelaya, Jhodi-Ann B. Dean, Caleb N. Chadwick, Don E. Davis, Joshua N. Hook, and Jesse Owen. 2023. "Multiple microaggressions and therapy outcomes: The indirect effects of cultural humility and working alliance with Black, Indigenous, women of color clients." *Professional Psychology: Research and Practice, Vol. 54*, no. 2, pp. 115-24. https://doi.org/10.1037/pro0000497

Denson, Nida, Katalin Szelényi, and Kate Bresonis. 2018. "Correlates of work-life integration for faculty across racial/ethnic groups." *Research in Higher Education, Vol. 59*, no. 2, pp. 226-47. https://doi.org/10.1007/s11162-017-9464-0

Dias, Felipe A., Joseph Chance, and Arianna Buchanan. 2020. "The motherhood penalty and the fatherhood premium in employment during COVID-19: Evidence from the United States." *Research in Social Stratification and Mobility, Vol. 69, 100542*, pp. 1-4. https://doi.org/10.1016/j.rssm.2020.100542

Diehl, Amy B., and Leanne M. Dzubinski. 2016. "Making the invisible visible: A cross-sector analysis of gender-based leadership barriers." *Human Resource Development Quarterly, Vol. 27*, no. 2, pp. 181-206. https://doi.org/10.1002/hrdq.21248

Dietz, Mary G. 2003. "Current controversies in feminist theory." *Annual Review of Political Science, Vol. 6*, no. 1, pp. 399-431. https://doi.org/10.1146/annurev.polisci.6.121901.085635

Donovan, Josephine. 1992. *Feminist Theory: The Intellectual Traditions of American Feminism*. Burns and Oates.

Durand, Michelle Kuntz, Ivonete Teresinha Schülter Buss Heidemann, Pamela Camila Fernandes Rumor, Carine Vendruscolo, Aline Megumi Arakawa Belaunde, and Jeane Barros de Souza. 2021. "Possibilities and challenges for female empowerment: perspectives of women in social vulnerability." *Anna Nery School, Vol. 25*, no. 5, pp. 1-8. https://doi.org/10.1590/2177-9465-ean-2020-0524

Eagleton, Mary. 1986. *Feminist Literary Theory: A Reader*. Wiley-Blackwell.

Eagly, Alice H., and Carli, Linda L. 2019. Women and the labyrinth of leadership. In *HBR's 10 Must Reads on Women and Leadership*. Cambridge, MA: Harvard Business Review, pp. 1-20.

Eagly, Alice H., and Madeline E. Heilman. 2016. "Gender and leadership: Introduction to the special issue." *Leadership Quarterly, Vol. 27*, no. 3, pp. 349-53. https://doi.org/10.1016/j.leaqua.2016.04.002

Eisele, Heather, and Jane E. Stake. 2008. "The differential relationship of feminist attitudes and feminist identity to self-efficacy." *Psychology of Women Quarterly, Vol. 32*, no. 3, pp. 233-44. https://doi.org/10.1111/j.1471-6402.2008.00432.x

Ellemers, Naomi. 2014. "Women at work: How organizational features impact career development." *Policy Insights from the Behavioral and Brain Sciences, Vol. 1*, no. 1, pp. 46-54. https://doi.org/10.1177/2372732214549327

Ellemers, Naomi, Floor Rink, Belle Derks, and Michelle K. Ryan. 2012. "Women in high places: When and why promoting women into top positions can harm them individually or as a group (and how to prevent this)." *Research in Organizational Behavior, Vol. 32*, pp. 163-87. https://doi.org/10.1016/j.riob.2012.10.003

Everly, George S., and Amy B. Athey. 2023. *Leading Beyond Crisis: The Five Pillars of Transformative Resilient Leadership*. American Psychological Association.

Falconi, April M., Ann M. Weber, Mark R. Cullen, Marcia L. Stefanick, Yvonne L. Michael, and Gary L. Darmstadt. 2020. "Shifts in women's paid employment participation during the World War II era and later life health." *Journal of Adolescent Health, Vol. 66*, no. 1, pp. S42-50. https://doi.org/10.1016/j.jadohealth.2019.10.005

Fedi, Angelina, and Chiara Rollero. 2016. "If stigmatized, self-esteem is not enough: Effects of sexism, self-esteem and social identity on leadership aspiration." *Europe's Journal of Psychology, Vol. 12*, no. 4, pp. 533-49. https://doi.org/10.5964/ejop.v12i4.984

Feng, Zhiyu, and Krishna Savani. 2020. "COVID-19 created a gender gap in perceived work productivity and job satisfaction: implications for dual-career parents working from home." *Gender in Management: An International Journal, Vol. 35*, no. 7-8, pp. 719-36. https://doi.org/10.1108/gm-07-2020-0202

Fischbach, Andrea, Philipp W. Lichtenthaler, and Nina Horstmann. 2015. "Leadership and gender stereotyping of emotions." *Journal of Personnel Psychology, Vol. 14*, no. 3, pp. 153-62. https://doi.org/10.1027/1866-5888/a000136

Fisher, Alexandra N., and Michelle K. Ryan. 2021. "Gender inequalities during COVID-19." *Group Processes and Intergroup Relations, Vol. 24*, no. 2, pp. 237-45. https://doi.org/10.1177/1368430220984248

Fitzsimmons, Terrance W., and Victor J. Callan. 2016. "Applying a capital perspective to explain continued gender inequality in the C-suite." *Leadership Quarterly, Vol. 27*, no. 3, pp. 354-70. https://doi.org/10.1016/j.leaqua.2015.11.003

Forbes, Lisa K., Margaret R. Lamar, and Rachel S. Bornstein. 2020. "Working mothers' experiences in an intensive mothering culture: A phenomenological qualitative study." *Journal of Feminist Family Therapy, Vol. 3,* no. 3, pp. 270-94. https://doi.org/10.1080/08952833.2020.1798200

Friedan, Betty. 2001. *The Feminine Mystique*. W. W. Norton and Company.

Friedman, Betty. 2005. "The soldier speaks": Yank coverage of women and wartime work." *American Journalism, Vol. 22*, no. 2, pp. 63-82. https://doi.org/10.1080/08821127.2005.10677640

Fritz, Claudia, and Daan Van Knippenberg. 2018. "Gender and leadership aspiration: The impact of work–life initiatives." *Human Resource Management, Vol. 57*, no. 4, pp. 855-68. https://doi.org/10.1002/hrm.21875

Ganesh, Shiv, and Heather M. Zoller. 2012. "Dialogue, activism, and democratic social change." *Communication Theory, Vol. 22*, no. 1, pp. 66-91. https://doi.org/10.1111/j.1468-2885.2011.01396.x

Garcini, Luz M., Melanie M. Domenech Rodríguez, Alfonso Mercado, and Manuel Paris. 2020. "A tale of two crises: The compounded effect of COVID-19 and anti-immigration policy in the United States." *Psychological Trauma: Theory, Research, Practice, and Policy, Vol. 12*, no. S1, pp. S230-32. https://doi.org/10.1037/tra0000775

Glass, Christy, and Alison Cook. 2018. "Do women leaders promote positive change? Analyzing the effect of gender on business practices and diversity initiatives." *Human Resource Management, Vol. 57*, no. 4, pp. 823-37. https://doi.org/10.1002/hrm.21838

Goldin, Claudia D. 1991. "The role of World War II in the rise of women's employment." *American Economic Review, Vol. 81*, no. 4, pp. 741-56. https://www.jstor.org/stable/2006640

Goler, Lori, Janelle Gale, Brynn Harrington, and Adam Grant. 2018. The 3 things employees really want: Career, community, cause. *Harvard Business Review, Vol. 20*. https://hbr.org/2018/02/people-want-3-things-from-work-but-most-companies-are-built-around-only-one?utm_medium=socialandutm_campaign=hbrandutm_source=LinkedInandtpcc=orgsocial_edit

Goode, Rachel W., Kevan Schultz, David Halpern, Sarah Godoy, Trenette Clark Goings, and Mimi Chapman. 2022. "Necessary, yet mistreated: The lived experiences of Black women essential workers in dual pandemics of racism and COVID-19." *Journal of Ethnic and Cultural Diversity in Social Work, Vol. 31*, no. 1, pp. 263-74. https://doi.org/10.1080/15313204.2022.2081642

Gray, Michael P., and Karen M. O'Brien. 2007. "Advancing the assessment of women's career choices: The career aspiration scale." *Journal of Career Assessment, Vol. 15*, no. 3, pp. 317-37. https://doi.org/10.1177/1069072707301211

Gurung, Lina. 2020. "Feminist standpoint theory: Conceptualization and utility." *Dhaulagiri Journal of Sociology and Anthropology, Vol. 14*, pp. 106-15. https://doi.org/10.3126/dsaj.v14i0.27357

Gutiérrez, Lorraine Margot, and Edith Anne Lewis. 1999. *Empowering Women of Color*. Columbia University Press.

Guzmán, Isabel Molina, and Angharad N. Valdivia. 2004. "Brain, brow, and booty: Latina iconicity in U.S. popular culture." *Communication Review, Vol. 7*, no. 2, pp. 205-21. https://doi.org/10.1080/10714420490448723

Hall, Martha L., Belinda T. Orzada, and Dilia Lopez-Gydosh. 2015. "American women's wartime dress: Sociocultural ambiguity regarding women's roles during World War II." *Journal of American Culture, Vol. 38*, no. 3, pp. 232-42. https://doi.org/10.1111/jacc.12357

Halliday, Cynthia Saldanha, Samantha C. Paustian-Underdahl, and Stav Fainshmidt. 2020. "Women on boards of directors: A meta-analytic examination of the roles of organizational leadership and national context for gender equality." *Journal of Business and Psychology, Vol. 36*, pp. 173-91. https://doi.org/10.1007/s10869-019-09679-y

Harvard Business Review Analytic Services. 2021. *Creating a Culture of Diversity, Equity, and Inclusion: Real Progress Requires Sustained Commitment*. Harvard Business School Publishing. https://hbr.org/sponsored/2021/09/creating-a-culture-of-diversity-equity-and-inclusion

Hays, Sharon. 1996. "The cultural contradictions of motherhood." *Social Forces, Vol. 76*, no. 3, pp. 1159-60. https://doi.org/10.2307/3005721

Hegewisch, Ariane, and Eve Mefferd. 2021. "A slow climb back from the "she-cession": High jobs deficit in childcare and school sectors continues (Q096)." *Institute for Women's Policy Research*, p. 111. https://files.eric.ed.gov/fulltext/ED614762.pdf

Hewlett, Sylvia A. 2008. Off-ramps and on-ramps: Keeping talented women on the road to success. *Human Resource Management International Digest, Vol. 16*, no. 2. https://doi.org/10.1108/hrmid.2008.04416bae.003

Hideg, Ivona, and Winny Shen. 2019. "Why still so few? A theoretical model of the role of benevolent sexism and career support in the continued underrepresentation of women in leadership positions." *Journal of Leadership and Organizational Studies, Vol. 26*, no. 3, pp. 287-303. https://doi.org/10.1177/1548051819849006

Hill-Jarrett, Tanisha G., and Martinque K. Jones. 2021. "Gendered racism and subjective cognitive complaints among older Black women: The role of depression and coping." *Clinical Neuropsychologist, Vol. 36*, no. 2, pp. 479-502. https://doi.org/10.1080/13854046.2021.1923804

Hoyt, Crystal L., and Susan E. Murphy. 2016. "Managing to clear the air: Stereotype threat, women, and leadership." *Leadership Quarterly, Vol. 27*, no. 3, pp. 387-99. https://doi.org/10.1016/j.leaqua.2015.11.002

Humm, Maggie. 1995. *The Dictionary of Feminist Theory* (2nd ed.). Ohio State University Press.

Hymowitz, Carol, and Michaele Weissman. 1978. *A History of Women in America: From Founding Mothers to Feminists—How Women Shaped the Life and Culture of America*. Bantam Books.

Ibarra, Herminia, Robin Ely, and Deborah Kolb. 2019. Women rising: The unseen barriers. In *HBR's 10 Must Reads on Women and Leadership*. Cambridge, MA: Harvard Business Review, pp. 39-50.

Jackson, Stevi, and Jackie Jones. 1998. *Contemporary feminist theories*. NYU Press.

Johnson, Natasha N. 2021. "Balancing race, gender, and responsibility: Conversations with four Black women in educational leadership in the United States of America." *Educational Management Administration and Leadership, Vol. 49*, no. 4, pp. 624-43. https://doi.org/10.1177/1741143221991839

Jones, Reece. 2021. *White Borders: The History of Race and Immigration in the United States from Chinese Exclusion to the Border Wall*. Beacon Press.

Kaiser, Robert B., and Wanda T. Wallace. 2016. "Gender bias and substantive differences in ratings of leadership behavior: Toward a new narrative." *Consulting Psychology Journal: Practice and Research, Vol. 68*, no. 1, pp. 72-98. https://doi.org/10.1037/cpb0000059

Kalysh, Kateryna, Carol T. Kulik, and Sanjeewa Perera. 2016. "Help or hindrance? Work-life practices and women in management." *Leadership Quarterly, Vol. 27*, no. 3, pp. 504-18. https://doi.org/10.1016/j.leaqua.2015.12.009

Kerr, Margaret L., Hannah F. Rasmussen, Kerrie A. Fanning, and Sarah M. Braaten. 2021. "Parenting during COVID-19: A study of parents' experiences across gender and income levels." *Family Relations, Vol. 70*, no. 5, pp. 1327-42. https://doi.org/10.1111/fare.12571

Khanmalek, Tala. 2020. How we get free: Black feminism and the Combahee River collective. *Oral History Review, Vol. 47*, no. 1, pp. 139-40. https://doi.org/10.1080/00940798.2019.1704155

Kim, Marlene. 2020. "Intersectionality and gendered racism in the United States: A new theoretical framework." *Review of Radical Political Economics, Vol. 52*, no. 4, pp. 616-25. https://doi.org/10.1177/0486613420926299

Kim, Tatyana, and Huang Gege. 2021. "She cession"—The decline in the status of women due to COVID-19." *4th Global Congress on Contemporary Sciences and Advancements, Vol. 4*, pp. 67-9. https://papers.econferenceglobe.com/index.php/ecg/article/view/371/368

Kleinman, Sherryl. 2007. *Feminist Fieldwork Analysis*. Sage.

Kolmar, Wendy K., and Frances Bartkowski. 2000. A black feminist statement. *Feminist Theory: A Reader*. McGraw-Hill Humanities, Social Sciences and World Languages, pp 272-277.

Ladge, Jamie J., Beth K. Humberd, and Kimberly A. Eddleston. 2018. "Retaining professionally employed new mothers: The importance of maternal confidence and workplace support to their intent to stay." *Human Resource Management, Vol. 57*, no. 4, pp. 883-900. https://doi.org/10.1002/hrm.21889

Landivar, Liana Christin, Leah Ruppanner, William J. Scarborough, and Caitlyn Collins. 2020. "Early signs indicate that COVID-19 is exacerbating gender inequality in the labor force." *Socius: Sociological Research for a Dynamic World, Vol. 6*, pp. 1–3. https://doi.org/10.1177/2378023120947997

Lee, Jaeeun, and Jennifer L. Wessel. 2022. "Is feminist identity beneficial for women's career aspirations? Examining feminist identity profiles." *Psychology of Women Quarterly, Vol. 46*, no. 1, pp. 27–49. https://doi.org/10.1177/03616843211055445

Lee, Minsun, Kristin Kim-Martin, Kimberly Molfetto, Kalya Castillo, Jessica L. Elliott, Yubelky Rodriguez, and Christopher M. Thompson. 2022. "Bicultural Asian American women's experience of gender roles across cultural contexts: A narrative inquiry." *Qualitative Psychology, Vol. 9*, no. 1, pp. 62–80. https://doi.org/10.1037/qup0000214

Leicht, Carola, Małgorzata A. Gocłowska, Jolien A. Van Breen, Soledad de Lemus, and Georgina Randsley de Moura. 2017. "Counter-stereotypes and feminism promote leadership aspirations in highly identified women." *Frontiers in Psychology, Vol. 8*, p. 883. https://doi.org/10.3389/fpsyg.2017.00883

Lewis, Jioni A. 2023. "Contributions of Black psychology scholars to models of racism and health: Applying intersectionality to center Black women." *American Psychologist, Vol. 78*, no. 4, pp. 576-88. https://doi.org/10.1037/amp0001141

Lewis, Jioni A., and Helen A. Neville. 2015. "Construction and initial validation of the gendered racial microaggressions scale for Black women." *Journal of Counseling Psychology, Vol. 62*, no. 2, pp. 289-302. https://doi.org/10.1037/cou0000062

Li, Peggy. 2013. "Hitting the ceiling: An examination of the barriers to success for Asian American women." *Berkeley Journal of Gender, Law and Justice, Vol. 29*, p. 140. https://doi.org/10.2139/ssrn.2254711

Lowman, Rodney L. 2005. "Importance of diagnosis in organizational assessment: Harry Levinson's contributions." *Psychologist-Manager Journal, Vol. 8*, no. 1, pp. 17–28. https://doi.org/10.1207/s15503461tpmj0801_3

Lyness, Karen S., and Angela R. Grotto. 2018. "Women and leadership in the United States: Are we closing the gender gap?" *Annual Review of Organizational Psychology and Organizational Behavior, Vol. 5*, no. 1, pp. 227-65. https://doi.org/10.1146/annurev-orgpsych-032117-104739

Lyttelton, Thomas, Emma Zang, and Kelly Musick. 2021. "Telecommuting and gender inequalities in parents' paid and unpaid work before and during the COVID-19 pandemic." *Journal of Marriage and Family, Vol. 84*, no. 1, pp. 230-49. https://doi.org/10.1111/jomf.12810

Mahalingam, Ramaswami. 2007. "Essentialism, power, and the representation of social categories: A folk sociology perspective." *Human Development, Vol. 50*, no. 6, pp. 300-19. https://doi.org/10.1159/000109832

Marsh-McDonald, Crystale M., and Sybil Schroeder. 2012. "Women in transition: A qualitative analysis of definitions of poverty and success." *Qualitative Report, Vol. 17*, no. 46, pp. 1-22. https://doi.org/10.46743/2160-3715/2012.1707

Maslow, Abraham H. 1943. "A theory of human motivation." *Psychological Review, Vol. 50*, no. 4, pp. 370-96. https://doi.org/10.1037/h0054346

Maslow, Abraham H. 1998. *Maslow on Management.* John Wiley and Sons.

Maslow, Abraham H. 2000. *The Maslow Business Reader* (D. C. Stephens, Ed.). John Wiley and Sons.

Maslow, Abraham H. 2013. *A Theory of Human Motivation.* Simon and Schuster.

Massey, Douglas S. 2020. "Immigration policy mismatches and counterproductive outcomes: Unauthorized migration to the U.S. in two eras." *Comparative Migration Studies, Vol. 8*, no. 1, pp. 1-27. https://doi.org/10.1186/s40878-020-00181-6

McCammon, Holly J., Karen E. Campbell, Ellen M. Granberg, and Christine Mowery. 2001. "How movements win: Gendered opportunity structures and U.S. women's suffrage movements, 1866 to 1919." *American Sociological Review, Vol. 66*, no. 1, pp. 49-70. https://doi.org/10.2307/2657393

McClain, Linda C. 2020. "What becomes a legendary constitutional campaign most? Marking the nineteenth amendment at one hundred." *Boston University Law Review, Vol. 100*, no. 5, p. 1753. https://scholarship.law.bu.edu/cgi/viewcontent.cgi?article=2021andcontext=faculty_scholarship

Meisenbach, Rebecca J. 2009. "The female breadwinner: Phenomenological experience and gendered identity in work/family spaces." *Sex Roles, Vol. 62*, no. 1-2, pp. 2-19. https://doi.org/10.1007/s11199-009-9714-5

Moe, Karine S., and Dianna Shandy. 2010. *Glass Ceilings and 100-Hour Couples: What the Opt-Out Phenomenon Can Teach Us About Work and Family.* University of Georgia Press.

Möhring, Katja, Maximiliane Reifenscheid, Andreas Weiland, and Klara Kuhn. 2021. "Is the recession a 'shecession'? Gender inequality in the employment effects of the COVID-19 pandemic in Germany." *SocArXiv*, pp. 1-41. https://doi.org/10.31235/osf.io/tzma5

Moody, Anahvia T., Jioni A. Lewis, and Gina P. Owens. 2022. "Gendered racism, coping, and traumatic stress among Black women: The moderating roles of the strong Black woman schema and womanist attitudes." *Psychology of Women Quarterly, Vol. 47*, no. 2, pp. 197-212. https://doi.org/10.1177/03616843221143752

Moya, Paula M. L. 2015. "Chicana feminism and postmodernist theory." *Signs, Vol. 26*, no. 2, pp. 441-83. https://doi.org/10.1086/495600

Mukkamala, Shruti, and Karen L. Suyemoto. 2018. "Racialized sexism/sexualized racism: A multimethod study of intersectional experiences of discrimination for Asian American women." *Asian American Journal of Psychology, Vol. 9*, no. 1, pp. 32-46. https://doi.org/10.1037/aap0000104

Muñoz, Cecillia. 2020. *More Than Ready: Be Strong and Be You … And Other Lessons for Women of Color on the Rise.* Seal Press.

Nasar, Saima. 2020. "Remembering Edward Colston: Histories of slavery, memory, and Black globality." *Women's History Review, Vol. 29*, no. 7, pp. 1218-25. https://doi.org/10.1080/09612025.2020.1812815

Nicholson, Linda J. 1997. *The Second Wave: A Reader in Feminist Theory.* Psychology Press.

Northouse, Peter G. 2018. *Leadership: Theory and Practice* (8th ed.). Sage.

O'Conner, Sandra D. 1996. "The history of the women's suffrage movement." *Vanderbilt Law Review, Vol. 49*, no. 3, pp. 657-76. https://assets.ctfassets.net/t9guaos1bsy4/LeIG3MO0J63iF1ky4LlwG/c571d3780b74389dda9974f4bb8b6a25/49VandLRev657-nofirst.pdf

Oleschuk, Merin. 2020. "Gender equity considerations for tenure and promotion during COVID-19." *Canadian Review of Sociology/Revue Canadienne de Sociologie, Vol. 57*, no. 3, pp. 502-15. https://doi.org/10.1111/cars.12295

Organization of American Historians. 2019. "Interchange: Women's suffrage, the nineteenth amendment, and the right to vote." *Journal of American History, Vol. 106*, no. 3, pp. 662-94. https://doi.org/10.1093/jahist/jaz506

Ozkazanc-Pan, Banu, and Alison Pullen. 2020. "Reimagining value: A feminist commentary in the midst of the COVID-19 pandemic." *Gender, Work and Organization, Vol. 28*, no. 1, pp. 1-7. https://doi.org/10.1111/gwao.12591

Pandey, Sonal. 2021. "COVID-19 and recession or shecession?" *American International Journal of Business Management, Vol. 4*, no. 4, pp. 5-13. https://www.aijbm.com/wp-content/uploads/2021/04/B440513.pdf

Paustian-Underdahl, Samantha C., Eden B. King, Steven G. Rogelberg, Clara Kulich, and William A. Gentry. 2017. "Perceptions of supervisor support: Resolving paradoxical patterns across gender and race." *Journal of Occupational and Organizational Psychology, Vol. 90*, no. 3, pp. 436-57. https://doi.org/10.1111/joop.12179

Paweenawat, Sasiwimon Warunsiri, and Lusi Liao. 2021. "A 'she-session'? The impact of COVID-19 on the labour market in Thailand." *ERIA Discussion Paper Series, Vol. 378*, pp. 1-36. https://www.eria.org/uploads/media/discussion-papers/ERIA-Research-on-COVID-19/A-'She-session'_The-Impact-of-COVID-19-on-the-Labour-Market-in-Thailand.pdf

Phipps, Simone T.A., and Leon C. Prieto. 2020. "Leaning in: A historical perspective on influencing women's leadership." *Journal of Business Ethics, Vol. 173*, no. 2, pp. 245-59. https://doi.org/10.1007/s10551-020-04566-6

Pilcher, Jane, and Imelda Whelehan. 2004. *50 Key Concepts in Gender studies.* Sage.

Pitt-Catsouphes, Marcie, Tay McNamara, Ragan Decker, Kathleen Christensen, Samuel Bradley, and Kerri Nelson. 2022. *The National Study of Workplace Equity.* SHRM.org. https://www.shrm.org/hr-today/trends-and-forecasting/research-and-surveys/pages/the-national-study-of-workplace-equity.aspx

Potter, David. 2021. *Disruption: Why Things Change.* Oxford University Press.

Power, Kate. 2020. "The COVID-19 pandemic has increased the care burden of women and families." *Sustainability: Science, Practice and Policy, Vol. 16*, no. 1, pp. 67-73. https://doi.org/10.1080/15487733.2020.1776561

Racine, Nicole, Rachel Eirich, Jessica Cooke, Jenney Zhu, Paolo Pador, Nicole Dunnewold, and Sheri Madigan. 2021. "When the bough breaks: A systematic review and meta-analysis of mental health symptoms in mothers of young children during the COVID-19 pandemic." *Infant Mental Health Journal, Vol. 43*, no. 1, pp. 36-54. https://doi.org/10.1002/imhj.21959

Randsley de Moura, Georgina, Carola Leicht, Ana C. Leite, Richard J. Crisp, and Małgorzata A. Gocłowska. 2018. "Leadership diversity: Effects of counterstereotypical thinking on the support for women leaders under uncertainty." *Journal of Social Issues, Vol. 74*, no. 1, pp. 165-83. https://doi.org/10.1111/josi.12262

Redfern, Bernice. 1989. *Women of Color in the United States: A Guide to the Literature.* Garland Publishing

Ridgeway, Cecillia L. 2014. "Why status matters for inequality." *American Sociological Review, Vol. 79*, no. 1, pp. 1-16. https://doi.org/10.1177/0003122413515997

Rincón Diez, Virginia, Miguel González, and Karle Barrero. 2017. "Women and leadership: Gender barriers to senior management positions." *Intangible Capital, Vol. 13*, no. 2, pp. 319-86. https://doi.org/10.3926/ic.889

Robnett, Rachael D., and Kristin J. Anderson. 2017. "Feminist identity among women and men from four ethnic groups." *Cultural Diversity and Ethnic Minority Psychology, Vol. 23*, no. 1, pp. 134-42. https://doi.org/10.1037/cdp0000095

Rojas, Maythee. 2009. *Women of Color and Feminism.* Seal Press.

Rosette, Ashleigh Shelby, Christy Zhou Koval, Anyi Ma, and Robert Livingston. 2016. "Race matters for women leaders: Intersectional effects on agentic deficiencies and penalties." *Leadership Quarterly, Vol. 27*, no. 3, pp. 429-45. https://doi.org/10.1016/j.leaqua.2016.01.008

Ryan, Michelle K., S. Alexander Haslam, Thekla Morgenroth, Floor Rink, Janka Stoker, and Kim Peters. 2016. "Getting on top of the glass cliff: Reviewing a decade of evidence, explanations, and impact." *Leadership Quarterly, Vol. 27*, no. 3, pp. 446-55. https://doi.org/10.1016/j.leaqua.2015.10.008

Seubert, Christian, Lisa Hopfgartner, and Jürgen Glaser. 2021. "Living wages, decent work, and need satisfaction: An integrated perspective." *European Journal of Work and Organizational Psychology, Vol. 30*, no. 6, pp. 808-23. https://doi.org/10.1080/1359432x.2021.1966094

Shim, Yerin, Bryan J. Dik, and James H. Banning. 2022. "Experiencing meaningful work as a lower socioeconomic status worker: An interpretative phenomenological analysis." *Journal of Career Assessment, Vol. 30*, no. 4, pp. 697-718. https://doi.org/10.1177/10690727221074891

Shockley, Kristen M., Malissa A. Clark, Hope Dodd, and Eden B. King. 2021. "Work-family strategies during COVID-19: Examining gender dynamics among dual-earner couples with young children." *Journal of Applied Psychology, Vol. 106*, no. 1, pp. 15-28. https://doi.org/10.1037/apl0000857

SHRM. 2021. "The culture effect: Why a positive workplace culture is the new currency." Society for Human Resource Management. https://www.shrm.org/hr-today/trends-and-forecasting/research-and-surveys/pages/2021-workplace-culture-report.aspx

Simmons, Sara V., and Robert M. Yawson. 2022. "Developing leaders for disruptive change: An inclusive leadership approach." *Advances in Developing Human Resources, Vol. 24*, no. 4, pp. 242-62. https://doi.org/10.1177/15234223221114359

Smith, Barbara. 2000. *Home Girls: A Black Feminist Anthology.* Rutgers University Press.

Snaebjornsson, Inga Minelgaite, Ingi Runar Edvardsson, Vilma Zydziunaite, and Vlad Vaiman. 2015. "Cross-cultural leadership." *SAGE Open, Vol. 5*, no. 2, pp. 1-8. https://doi.org/10.1177/2158244015579727

Snyder, Claire. 2008. "What is third-wave feminism? A new directions essay." *Signs: Journal of Women in Culture and Society, Vol. 34*, no. 1, pp. 175-96. https://doi.org/10.1086/588436

Soto, DeFrabcesci. 2020. "America's recovery from the 2020 "Shecession": Building a female future of childcare and work." In *LBJ School of Public Affairs Research*. YWCA. https://doi.org/10.26153/tsw/10385

Spates, Kamesha, Na'Tasha Evans, Tierra Akilah James, and Karen Martinez. 2020. "Gendered racism in the lives of Black women: A qualitative exploration." *Journal of Black Psychology, Vol. 46*, no. 8, pp. 583-606. https://doi.org/10.1177/0095798420962257

Sue, Derald Wing, Christina M. Capodilupo, Gina C. Torino, Jennifer M. Bucceri, Aisha Holder, Kevin L. Nadal, and Marta Esquilin. 2007. "Racial microaggressions in everyday life: Implications for clinical practice." *American Psychologist, Vol. 62*, no. 4, pp. 271-86. https://doi.org/10.1037/0003-066x.62.4.271

Swanson, Richard A. 2022. *Foundations of Human Resource Development* (3rd ed.). Berrett-Koehler.

Szymanski, Dawn M., and Destin N. Stewart. 2010. "Racism and sexism as correlates of African American women's psychological distress." *Sex Roles, Vol. 63*, no. 3-4, pp. 226-38. https://doi.org/10.1007/s11199-010-9788-0

Taormina, Robert J., and Jennifer H. Gao. 2013. "Maslow and the motivation hierarchy: Measuring satisfaction of the needs." *American Journal of Psychology, Vol. 126*, no. 2, pp. 155-77. https://doi.org/10.5406/amerjpsyc.126.2.0155

Taramundi, Dolores M. 2016. "Gender machineries vs. feminist movements? Collective political subjectivity in the time of passive revolution." *Gender and Education, Vol. 28*, no. 3, pp. 372-85. https://doi.org/10.1080/09540253.2016.1169253

Taylor, Zoe E., Keisha Bailey, Fabiola Herrera, Nayantara Nair, and Abrianna Adams. 2022. "Strengths of the heart: Stressors, gratitude, and mental health in single mothers during the COVID-19 pandemic." *Journal of Family Psychology, Vol. 36*, no. 3, pp. 346-57. https://doi.org/10.1037/fam0000928

Turner, Lynn H., Daradirek Ekachai, and Karen Slattery. 2022. "How working mothers juggle jobs and family during COVID-19: Communicating pathways to resilience." *Journal of Family Communication, Vol. 22*, no. 2, pp. 138-55. https://doi.org/10.1080/15267431.2022.2058510

US Bureau of Labor Statistics. 2021. *Women in the Labor Force: A Databook*. BLS Reports. Washington, DC. https://www.bls.gov/opub/reports/womens-databook/2020/home.htm

US Census Bureau. 2021. *Explore Census Data*. Washington, DC. https://data.census.gov/table?q=Povertyandtid=ACSST1Y2021.S1703

Vance, Michelle M., Jeannette M. Wade, Mervin Brandy, Jr., and Aiyana R. Webster. 2022. "Contextualizing Black women's mental health in the twenty-first century: Gendered racism and suicide-related behavior." *Journal of Racial and Ethnic Health Disparities, Vol. 10*, no. 1, pp. 83-92. https://doi.org/10.1007/s40615-021-01198-y

Vermote, Branko, Joachim Waterschoot, Sofie Morbée, Jolene Van der Kaap-Deeder, Charlotte Schrooyen, Bart Soenens, Richard Ryan, and Maarten Vansteenkiste. 2021. "Do psychological needs play a role in times of uncertainty? Associations with well-being during the COVID-19 crisis." *Journal of Happiness Studies, Vol. 23*, no. 1, pp. 257-83. https://doi.org/10.1007/s10902-021-00398-x

Vial, Andrea C., Jaime L. Napier, and Victoria L. Brescoll. 2016. "A bed of thorns: Female leaders and the self-reinforcing cycle of illegitimacy." *Leadership Quarterly, Vol. 27*, no. 3, pp. 400-14. https://doi.org/10.1016/j.leaqua.2015.12.004

Volini, Erica, J. Schwartz, I. Roy, M. Hauptmann, Y. Van Durme, B. Denny, and J. Bersin. 2019. "Leadership for the 21st century: The intersection of the traditional and the new." *2019 Global Human Capital Trends (2019)*. https://www2.deloitte.com/content/dam/insights/us/articles/5136_HC-Trends-2019/DI_HC-Trends-2019.pdf

Waldron, Ingrid R. G. 2019. Archetypes of Black womanhood: Implications for mental health, coping, and help-seeking. In *ADVANCES in Mental Health and Addiction*. Springer International, pp. 21-38. https://doi.org/10.1007/978-3-030-26437-6_2

Warner, Leah R., and Stephanie A. Shields. 2018. *Intersectionality as a Framework for Theory and Research in Feminist Psychology.* Oxford University Press eBooks. https://doi.org/10.1093/oso/9780190658540.003.0002

Wayne, Julie H., Russell Matthews, Wayne Crawford, and Wendy J. Casper. 2020. "Predictors and processes of satisfaction with work-family integration: Examining the role of personal, work, and family resources and conflict and enrichment." *Human Resource Management, Vol. 59*, no. 1, pp. 25-42. https://doi.org/10.1002/hrm.21971

Westring, Alyssa Friede, and Ann Marie Ryan. 2011. "Anticipated work–family conflict: A construct investigation." *Journal of Vocational Behavior, Vol. 79*, no. 2, pp. 596-610. https://doi.org/10.1016/j.jvb.2011.02.004

Whiley, Lilith A., Hazel Sayer, and Marie Juanchich. 2021. "Motherhood and guilt in a pandemic: negotiating the "new" normal with a feminist identity." *Gender, Work and Organization, Vol. 28*, no. S2, pp. 612-19. https://doi.org/10.1111/gwao.12613

White, Aaronette M. 2006. "Racial and gender attitudes as predictors of feminist activism among self-identified African American feminists." *Journal of Black Psychology, Vol. 32*, no. 4, pp. 455-78. https://doi.org/10.1177/0095798406292469

Williams, Carmen Braun, and Marsha I. Wiggins. 2010. "Womanist spirituality as a response to the racism-sexism double bind in African American women." *Counseling and Values, Vol. 54*, no. 2, pp. 175-86. https://doi.org/10.1002/j.2161-007x.2010.tb00015.x

Wollstonecraft, Mary. 2019. *The Feminist Papers: A Vindication of the Rights of Women*. Gibbs Smith.

Women in the Workplace. 2023. LeanIn.org and McKinsey and Company. https://sgff-media.s3.amazonaws.com/sgff_r1eHetbDYb/Women+in+the+Workplace+2023_+Designed+Report.pdf

Yawson, Robert. 2020. "Strategic flexibility analysis of HRD research and practice post COVID-19 pandemic." *Human Resource Development International, Vol. 23*, no. 4, pp. 406-17. https://doi.org/10.1080/13678868.2020.1779169

Zanhour, Mona, and Dana M. Sumpter. 2022. "The entrenchment of the ideal worker norm during the COVID-19 pandemic: Evidence from working mothers in the United States." *Gender Work and Organization, Vol. 31*, no. 2, pp. 625-43. https://doi.org/10.1111/gwao.12885

Zhang, Shihang, Rolf Moeckel, Ana Tsui Moreno, Bin Shuai, and Jie Gao. 2020. "A work-life conflict perspective on telework." *Transportation Research Part A: Policy and Practice, Vol. 141*, pp. 51-68. https://doi.org/10.1016/j.tra.2020.09.007

Ziglar, Tom. 2021. *10 Leadership Virtues for Disruptive Times: Coaching Your Team Through Immense Change and Challenge*. Thomas Nelson.

XII. Health and Safety

Is Bigger Better? Safety Implications of Local Trucking Labor Market Concentration

KEVIN L. CONNER
University of Utah

Introduction

Trucking is a dangerous industry wherein the risks are borne both by truck drivers and the public. In 2020 in the United States, there were more than 100,000 crashes involving large trucks, resulting in more than 140,000 injuries among truck drivers and the public (FMCSA 2022b). Existing research shows that drivers receiving higher remuneration and those working for larger companies tend to be safer (Belzer and Sedo 2018; Cantor et al. 2016; Kudo and Belzer 2019a, 2019b). However, if larger companies have sufficient labor market power, they may pay drivers less. Because of this, it is unclear whether larger firms—and the accompanying increase in labor market concentration—enhance or diminish safety in the industry.

To examine these relationships, I assembled state-level data covering local (intrastate) trucking between 1998 and 2020. These data were a combination of crash and injury information from the Federal Motor Carrier Safety Administration's (FMCSA) Motor Carrier Management Information System Crash Files (MICMIS-Crash) and economic information from the US Census Bureau's County Business Patterns (CBP). Local trucking was examined because it has more clearly defined market boundaries than long-distance interstate trucking.

I estimated two generalized structural equation models (GSEM), one for accidents and the other for injuries. These outcomes were modeled as a function of both market concentration and driver remuneration, while accounting for market concentration's negative association with driver remuneration. This approach permits estimation of the net association of market concentration with accidents and injuries and can show to what extent this relationship is mediated by local driver pay.

In simultaneously examining market concentration and driver remuneration, this study melded research showing that bigger firms and higher driver pay tend to be related to better safety. While I found that concentrated local trucking labor markets tend to be safer, the accompanying decline in driver remuneration offsets this relationship. I also found that increased pay among long-distance truck drivers is associated with increased pay for local truck drivers. By this pathway, increased long-distance driver pay was significantly associated with fewer accidents and injuries for local drivers based in the same state.

Literature Review

Existing research shows that higher driver pay and benefits are associated with fewer accidents and moving violations (Faulkiner and Belzer 2019; Kudo and Belzer 2019b; Rodríguez et al. 2003, 2006; Ryley and Belzer 2023). Drivers working for larger employers also tend to be safer (Cantor et al. 2016). Examination of competition in the trucking industry has typically focused on output markets rather than labor markets (Belzer 2000; Dempsey 2003; Engel 1998). Developments in the monopsony literature demonstrate that employer labor market power is pervasive, even in apparently competitive labor markets (Ashenfelter et al. 2010). Together, this suggests that market concentration may have direct and indirect relationships with safety outcomes.

Crash Risks and Regulation

A major contributor to the workplace danger that truckers experience is the risk of accidents. Given the danger crashes pose to both truckers and the public, trucking is heavily regulated. Regulations limit the number of hours that a driver may continuously operate a truck and specify mandatory rest periods to lessen driver fatigue (FMCSA 2022a). Despite these regulations, drivers regularly report working more than the legal number of hours per week (Belzer and Sedo 2018).

Remuneration

Truck drivers in the United States are commonly paid piece rates based on the load or number of miles driven (Burks et al. 2010). Under piece rates, drivers may have some target level of earnings that they wish to achieve each working day—and they work until they achieve their target (Camerer et al. 1997). This target earnings behavior has been documented among commercial truck drivers (Belzer and Sedo 2018). If piece rates are low relative to trucker targets, truckers have an incentive to drive additional miles, but higher piece rates, or paying drivers for nondriving tasks, tend to reduce driving hours (Belzer and Sedo 2018; Kudo and Belzer 2019a).

Low driver pay and piece rates can lead to self-sweating, where drivers push "themselves to work or sacrifice more despite little or no financial gain and significant risk" (Viscelli 2016). Working long hours to meet their earnings targets, truckers may become fatigued—a well-documented accident risk (Bunn et al. 2005, 2019; Jovanis et al. 2011, 2011; Soliani et al. 2023; Torregroza-Vargas et al. 2014). Consistent with the target earnings understanding, Kudo and Belzer (2019b) showed greater driver remuneration was associated with fewer moving violations, a proxy measure for safety performance. Higher driver remuneration is also associated with fewer crashes (Rodríguez et al. 2003; Ryley and Belzer 2023). Higher compensation can also help retain older, more experienced drivers, who tend to have fewer crashes (Faulkiner and Belzer 2019; Rodríguez et al. 2006). In short, there is a well-established relationship between driver pay and safety in the trucking industry.

Competition

Prior to the passage of the Motor Carrier Act of 1935, the trucking industry was in a downward spiral of falling rates. This descent was fueled by a queue of new for-hire carriers ready to enter the industry (Dempsey 2003). The 1935 Motor Carrier Act sought to arrest this fall by moderating interfirm competition (Dempsey 2003). Between 1935 and 1980, the Interstate Commerce Commission regulated entry, exit, and the rates that carriers could charge on each route (Dempsey 2003). After 1980, this system of regulation waned with a deregulatory tilt in Interstate Commerce Commission decisions, the passage of the Motor Carrier Act of 1980, and ultimately, the dissolution of the Interstate Commerce Commission in 1995 (Belzer 2000).

In the wake of deregulation, trucking became substantially more competitive (Belzer 2000; Engel 1998). The number of registered for-hire carriers increased by more than sevenfold between the passage of the Motor Carrier Act in 1980 and 2007 (Burks et al. 2010). Intensified price competition among carriers placed downward pressure on driver remuneration (Belzer 2000). Although it occurred later, local trucking was also largely deregulated by the mid-1990s (Tang and Ma 2002; Trick and Peoples 2012). While segments with higher barriers to entry were somewhat insulated, deregulation intensified price competition and led to a fall in driver compensation (Belman and Monaco 2001; Engel 1998). Emphasizing this, the relatively less competitive less-than-truckload (LTL) segment tends to have a greater union presence, higher pay, better safety performance, and fewer crashes (Corsi et al. 2012).

Intense competition can place carriers in similarly precarious circumstances to those of their drivers. Carriers experiencing poor financial performance also tend to be less safe (Britto et al. 2010; Miller and Saldanha 2016). These studies suggest several potential causal mechanisms. Struggling carriers may pressure their drivers to work longer hours, pay their drivers less, and thus fail to retain more experienced drivers. Struggling carriers may skimp on other determinants of safety, such as vehicle maintenance. These carriers may also have fewer resources devoted to driver monitoring and training. Provided that carriers in less competitive markets have better financial performance, there is potential for a direct relationship between market concentration and safety. Moreover, larger firms, which would be more prevalent in more concentrated labor markets, tend to have better safety performance (Cantor et al. 2016). In short, both precarious remuneration and precarious profitability are associated with greater crash risks.

Labor Market Competition

Existing work on the trucking industry emphasizes the effects of intense price competition among carriers in output markets on labor standards. There has been relatively little investigation of the potential impacts of competition on labor markets. This is partly due to a traditional understanding of market power, which focuses on the number of buyers and sellers. A seller of a good or service has market power when there are few suppliers of similar goods and services. Likewise, an employer has power in the labor market when workers have few alternative employers or employers are acting in concert (Robinson 1969). However, exclusively focusing on the number of employers can understate the extent of employer labor market power. Even with many potential employers, workers may still stick with a job even when the wage rate falls below that of competing employers due to other limitations on worker mobility.

Developments in the modern monopsony literature emphasize job differentiation and search frictions (Manning 2003). If a labor market is a friction-free, perfectly competitive market, a worker will switch jobs if their pay falls even a penny below the market-clearing wage rate. Under these assumptions, the perfectly competitive employer faces a perfectly elastic labor supply curve at the firm level (Manning 2021). Modern monopsony-based empirical work usually estimates the firm-level elasticity of the labor supply, where deviations from perfect elasticity serve as indicators of employer labor market power.

By focusing on job differentiation and search frictions rather than the number of employers or differences in worker preferences, modern monopsony researchers have shown that employer market power is more pervasive than once believed (Ashenfelter et al. 2010). There is compelling evidence that positive firm-level elasticities of labor supply are common (Sokolova and Sorensen 2021). Given these developments, employer labor market power is a potential determinant of driver safety via its downward influence on driver remuneration.

Local and Long-Distance Connections and Distinctions

While interstate trucking makes up most of the industry, roughly 30% of drivers worked for local carriers in 2020 (US Census Bureau 2024). Interstate carriers operate across state lines and must register with the US Department of Transportation and comply with all federal regulations. Local carriers operate within a single state and must only satisfy the requirements of their respective state authorities, which often overlap with federal requirements.

Some common local trucking jobs include local deliveries to and from large transportation terminals, such as railyards and ports; deliveries to businesses on a regular route; last-mile drivers for LTL services; or other local parcel carriers. Regardless, local drivers tend to be home most nights, while interstate drivers must spend substantial time away from home. Because local trucking is by definition local, there may be fewer potential employers and labor market frictions associated with local work. Much of the research on driver compensation and safety focuses on interstate trucking, with only one study specifically focusing on local trucking (Ju and Belzer 2024).

I focused on local trucking because it is more plausibly composed of state-level labor markets, enabling a state-level analysis. A prospective truck driver selecting among interstate carriers likely has many more options than a similar driver selecting among only local carriers. From the more recent perspective of labor market frictions, a trucker in a local labor market may be less willing to leave a current job if they prefer certain aspects of the job, such as frequent home time. Thus, from both a traditional monopsony perspective and a more recent friction perspective, labor market concentration may matter more for local trucking than for long-distance trucking.

Still, these two labor markets are interrelated. Workers decide among competing employers based on wages and other job qualities. Local truckers may be willing to accept lower earnings in exchange for more home time or because of other aspects of local trucking employment. However, sufficiently high earnings may still draw some local drivers into the interstate sector. Because of this, it is important to account for local truckers' labor market alternatives of working for an interstate carrier.

Data

This section reviews each measure used in the analysis, its calculation, and the dataset from which it was gathered. All data are annual at the state level and cover the years 1998 through 2020. The two principal sources of data are the Census Bureau's County Business Patterns (CBP) and the Federal Motor Carrier Safety Administration's Motor Carrier Management Information System Crash Files (MCMIS-Crash).

County Business Patterns: Market Concentration, Average Pay, and Employment

Market concentration, average pay, and employment were all taken from the Census Bureau's CBP state-level files (US Census Bureau 2024). Employment in local trucking was taken directly from Census-published files for local general freight trucking (NAICS 484110) and local specialized freight trucking (NAICS 484220). These employment numbers were combined with total payroll figures in the CBP to calculate the average annual payroll per employee, henceforth called average driver pay. Identical calculations were done for long-distance trucking NAICS codes 484121 (general freight trucking, long-distance, truckload) and 484230 (specialized freight trucking, long-distance) to estimate the average annual long-distance driver pay in each state. For long-distance drivers, it is important to note that "state" is determined by employer location rather than a worker's residence.

These average pay figures include both legally required and fringe benefits. All nominal dollar values were adjusted to 2018 dollars using the Consumer Price Index for Urban Wage Earners and Clerical Workers (US BLS 2023a). These values are approximate because the CBP counts all workers, including nondrivers, at an establishment. Nonetheless, the US BLS reports that in 2022, 82% of all trucking industry employees were drivers, 9% were laborers, 5% were mechanics, and 4% were machine operators and supervisors (2024).

The most common measure of market concentration is the Hirsch–Herfindal Index (HHI). The Census does not publish the establishment-level figures necessary to calculate the HHI. Instead, it publishes employment and establishment counts by employer-size categories. I use these binned employment values to calculate a pseudo-HHI measure. A small number of these binned employment counts are censored due to a few observations. When an observation is censored, a range is given. In these cases, I selected the midpoint of the range. In later years, the Census used artificial noise rather than censoring. I used the noise-injected values without alteration.

Using the binned employment values, I calculated a pseudo-HHI using a procedure in Meiseblach et al. (2022). I assumed that every establishment in an employer-size bin employs the average number of workers for employers in that bin. The formula is as follows:

$$\widetilde{HHI} = k_1\mu_1^2 + k_2\mu_2^2 + \cdots k_K\mu_K^2$$

where K is the number of bins, k_i is the number of establishments in bin i, and μ_i is the average number of employees among establishments in bin i.

Motor Carrier Management Information System: Accident and Injury Frequency

Incident and injury frequencies were taken from the FMCSA's MCMIS-Crash files (FMCSA 2019). These files contain all crashes reported by registered carriers. A carrier must report a crash if it results in a vehicle being towed away, an injury requiring off-site treatment, or a fatality (*Code of Federal Regulations* 2024). All interstate carriers are required to register, and many states require their local carriers to register. Carriers hauling specialized freight such as hazardous materials may also be required to register, even if they operate in only a single state.

I filtered the MCMIS-Crash files for local trucking establishments and aggregated these data to the state level. The validity of these aggregate measures is dependent on the cooperation of each state. In the earlier years of these data, some states reported implausibly low numbers, suggesting a lack of commitment to federal reporting. Consequently, I removed observations if the case count increased by more than 25% in one year and remained high in all subsequent periods without similar changes in local trucking employment. Removing these data and excluding a small number of observations missing employment data, 64 observations were dropped out of an initial 1,137.

Freight Analysis Framework: Activity Level

State-industry level productivity in ton-miles was estimated using information from the Federal Highway Administration's Freight Analysis Framework (FAF) (2023). The FAF provides goods flow data for different transport modes. I extracted all freight flows for which the origin and destination states were identical. The FAF is produced every fifth year (1997, 2002, 2007, 2012, 2017, and 2022) with imputations for intervening years provided since 2017. In the interest of consistency, I produced an imputation for the intervening years using multiple imputation linear regression. Using observed ton-mile values, I assumed that ton-miles are a linear function of employment and a linear trend. Each state's imputation used only that state's observations.

Annual productivity data at this level of geographic and industry detail are not readily available. There is a large proportion of local trucking activity that remains unobserved in the FAF data. Figures for ton-miles per worker were implausibly low for many states, indicating substantial measurement error. These imputed values should be taken as a rough measure of productivity that relies on the assumption that measurement errors are similar across states. The lack of a more comprehensive measure of productivity in intrastate trucking is a weakness of this analysis.

Other Variables

Urban density was calculated based on road network characteristics contained in the Office of Highway Policy Information's Miles by Functional System data (FHA 2022). Using these data, I calculated the urban share of principle arterial road miles. State-level unemployment rates were retrieved from the Bureau of Labor Statistic's local area employment statistics (US BLS 2023b).

Methods

Generalized Structural Equation Model

This analysis used a generalized structural equation model (GSEM). These models have two capabilities that enable this analysis. First, GSEMs can model nonnormally distributed count outcome variables. Second, GSEMs make it possible to calculate the sum-total association of labor market concentration with safety outcomes and determine to what extent this association is diminished by labor market concentration's indirect relationship to safety via driver pay.

GSEMs combine the capabilities of traditional structural equation modeling with those of generalized linear modeling (Baum 2021; Muthén and Asparouhov 2015). GSEM models nonnormally distributed outcome variables by specifying different combinations of family and link functions. GSEM also allows for estimating mediated associations without discrete treatment levels, as required in many other mediation methods. Depending on the causal pathways, introducing a mediator as a control variable into a typical regression model may lead to discrepancies between the regression coefficient and the hypothesized effect being measured (Cinelli et al. 2022).

The typical way to control for unobserved differences would be to use a fixed-effects framework with indicator variables for each state. However, GSEMs cannot easily absorb the large number of state-level indicator variables required for this approach. Without many degrees of freedom, introducing numerous control variables typically leads to nonconvergence. Instead, I explicitly controlled for variation in each state's trucking industry with observed continuous variables. Consequently, the models used in this analysis neglected the panel nature of these data and do not account for unobserved time-invariant differences in each state's local trucking industry.

Model Specification

I estimated two similar generalized structural equation models. In each model, coefficients were simultaneously estimated for two interrelated outcomes: average local driver pay and a safety outcome (one model for accidents and one for injuries). Both models were structured such that market concentration has a direct relationship to the safety outcome and an indirect relationship via average driver pay. Local driver earnings are also a function of the earnings of long-distance drivers based in the same state. Additional variables controlled for time trends and for determinants of driver pay and safety outcomes. Figure 1 illustrates the final model specification for the GSEMs.

Figure 1. Generalized structural equation model specification

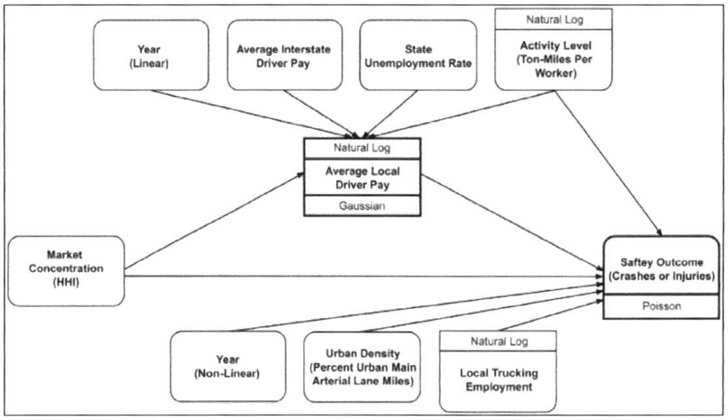

Bold boxes indicate outcome variables.

For the earnings estimation stage of the model, I used a Gaussian family and an identity link function because the earnings data are normally distributed after a natural log transformation. Crashes and injuries were counted variables that are neither zero-inflated nor overdispersed when aggregated to the state level. Therefore, the latter stages of the GSEMs used the Poisson family with a log-link function. Specification was guided by model comparisons using the Bayesian Information Criterion (BIC).

The variables included ton-miles per driver as a rough measure of productivity; state-level unemployment rates and interstate driver pay account for labor market conditions; and urban share of lane miles, which accounts for the urban density of state road networks. The unit used for market concentration in all models was a 100-point change in HHI. A 100-point policy is relevant, whereby regulators determine if a prospective merger would substantially lessen competition (DOJ-FTC 2023). Quantity variables were natural log–transformed only if doing so improved the BIC. In the preferred models, time was modeled linearly for the earnings estimation and nonlinearly for the safety outcomes, which resulted in the best model fit as measured by the BIC.

Count outcome models often contain an exposure variable to account for subjects being observed for different periods of time. An exposure variable is an independent variable wherein its increase should result in a directly proportional increase in the count variable (Cameron and Trivedi 2022). Exposure variables are log-transformed and have their coefficient constrained to 1, which enables interpretation in terms of incidence rate ratios. It is not unreasonable to suggest that employment in this model functions similarly to the time under observation in more traditional models. All else being equal, each additional driver creates additional opportunities for crashes and injuries. However, there may be systematic differences in injuries between states with different-sized local trucking industries.

To account for this possibility, employment was log-transformed, but the coefficient was left unconstrained. Linear hypothesis tests also indicate that the coefficients of the log of employment were statistically different from one in both models under a typical 95% confidence interval.

Mediated effects are the product of two normally distributed coefficients, which are not necessarily normally distributed, making typical confidence intervals inappropriate. Instead, I used bootstrapped standard errors, which do not impose a particular distribution on the parameter being tested (Albert 2012; Fairchild and McDaniel 2017).

Results

Over the time for which these data were collected, local trucking has become less concentrated, better remunerated, and safer in terms of injuries. The descriptive trends show that labor markets for local truck drivers have become less concentrated, albeit with a more recent increase in concentration. This fall in concentration coincides with increases in real local truck driver pay and a reduction in the pay premium attached to interstate work. While there is no clear trend in local trucking accidents, there has been a near continual decline in the severity of accidents as measured by injuries per crash incident. Percentiles in all subsequent figures are based on state-level observations and frequency weighted by local trucking employment.

Figure 2 shows local trucking labor market concentration over time using the quartile values of HHI. Rather than a clear trend, there seems to be a period of higher concentration, followed by a more recent period of lower concentration. Even in the more highly concentrated period, the market concentration remained quite low. The Department of Justice does not consider a market to be moderately concentrated until HHI exceeds 1000 (2023). In these data, a vanishingly small number of states have local trucking labor markets meeting this criterion.

Figure 2. Local trucking labor market concentration over time

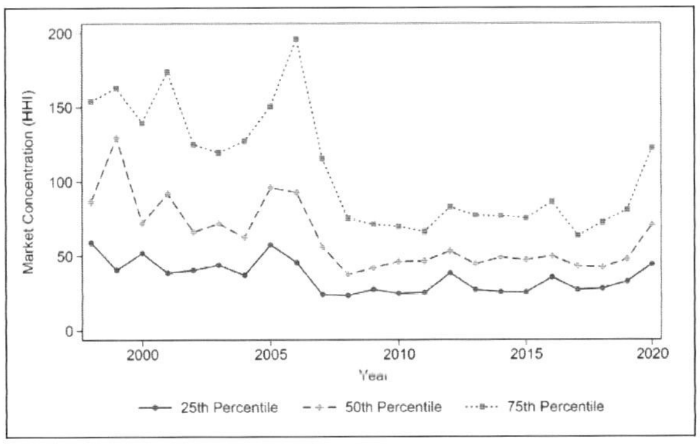

Figure 3 illustrates the average real driver compensation. The upper panel shows that local driver pay has increased over time. This rise in pay roughly coincides with the latter period of the lower labor market concentration shown in Figure 2. This pattern is preceded by a brief period (2005–2007) with the highest observed labor market concentration, which coincided with a period of high driver pay. The lower panel compares median local and interstate driver pay. The two values have converged, but there is still a persistent interstate pay premium. This narrowing may be due to changes in piece rates or changes in working hours since these total annual earnings do not account for changes in labor hours.

Figure 4 shows quartile values over time for the number of crash incidents per 1,000 workers and the number of injuries per crash incident. While there is no clear time trend for the number of crash incidents, there is a distinct decline in the number of injuries per incident.

Figure 3. Intrastate driver compensation over time

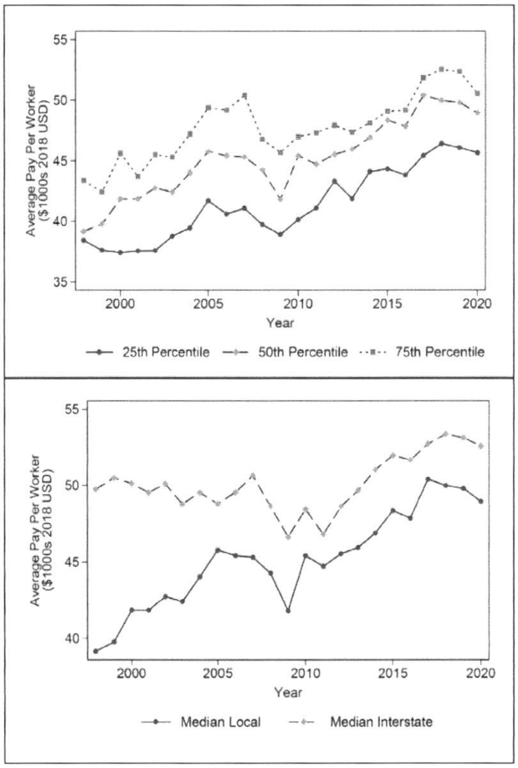

Figure 4. Local trucking safety outcomes over time

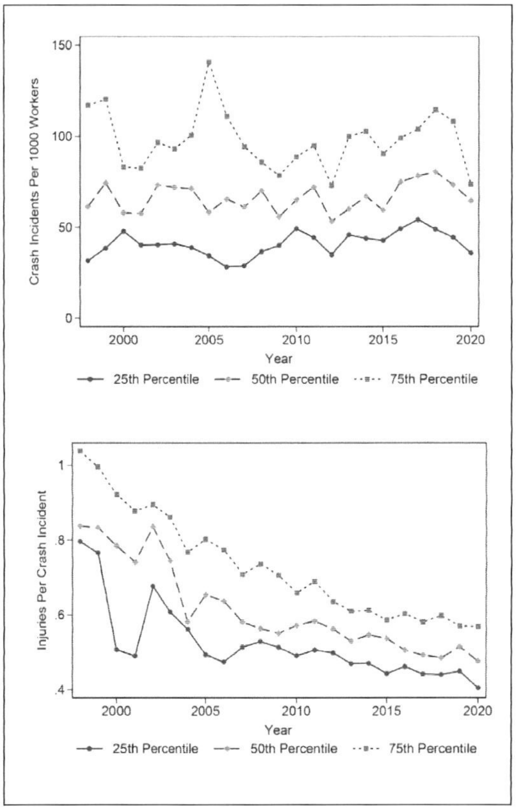

The descriptive trends show general improvements in earnings and safety in local trucking, but they do not demonstrate a conditional relationship between these variables and market concentration. To further unpack the complex relationships among these variables, I now turn to generalized structural equation models.

GSEM Estimates

Table 1 contains the GSEM results for the crash incident and injury models. The two models share an identical earnings estimation stage. The positive time trend in local driver pay that appeared descriptively also appears here. The state-level unemployment rate is of the expected sign but lacks statistical significance. For both models, there is a significant negative association between market concentration and average annual drivers' pay. Interstate drivers' pay, representing a close alternative to driving locally, was statistically significant and positively associated with increased local driver pay. Increased productivity was associated with lower driver pay.

The results are broadly similar for both safety outcome models. Because these models are count-outcome models, the coefficients can be exponentiated to provide incident rate ratios (IRRs). For injury results, the increased incident rate of injuries includes injuries to drivers as well as members of the public. A 1% increase in local driver pay is, on average, associated with approximately 0.057% fewer crashes and 0.072% fewer injuries.

Table 1. GSEM safety outcome model results

	Model 1: crash incidents	Model 2: injuries
	Log of local driver pay (annual)	
Market concentration (HHI)	–0.012***	–0.012***
Interstate driver pay (annual)	0.0160***	0.0160***
Log of activity level (ton-miles per worker)	–0.059***	–0.059***
State-level unemployment rate	–0.0001	–0.0001
Year (linear)	0.008***	0.008***
Constant	–13.56***	–13.56***
	Safety outcomes	
	Crash incidents	Injuries
Market concentration (HHI)	–0.057***	–0.072***
Log of local driver pay (annual)	–0.425***	–0.506***
Log of activity level (ton-miles per worker)	0.254***	0.291***
Log of local trucking employment	1.055***	0.993***
Urban density (percent urban main arterial lane miles)	0.009***	0.012***
Year (nonlinear); reference: 1998	†	†
Constant	–1.964***	–1.303***
N	1073	1073

†Coefficients for year indicator variables estimated but not reported. Nonlinear year indicator variables are jointly significant ($p < 0.001$).

*$p < 0.05$, **$p < 0.01$, ***$p < 0.001$.

Increased market concentration was directly associated with fewer crashes and injuries. A 100-point increase in HHI was directly associated with 5.4% fewer accidents and 6.9% fewer injuries. The sign of productivity level was also as expected, with additional ton-miles per worker being associated with higher rates of accident and injury. The road network also matters—with more urban road networks being associated with increased crashes and injuries.

Mediation

The sign and significance of the model coefficients suggest that driver earnings mediate the effect of market concentration on safety outcomes, but significance must be tested directly. Table 2 contains point estimates and p-values for the mediated indirect effects, direct effects, and total effect.

Table 2. HHI-local driver pay mediation significance tests

	HHI indirect effect (via local driver pay)		HHI direct effect		HHI total effect	
	Coeff.	**p-value**	**Coeff.**	**p-value**	**Coeff.**	**p-value**
Model 1: crash incidents	0.005	0.068	–0.057	0.000	–0.052	0.000
Model 2: injuries	0.006	0.079	–0.072	0.000	–0.066	0.000

Bootstrap standard errors use 3,000 replications.

The indirect association via earnings is small compared to the direct association of market concentration with fewer crashes and injuries, but the driver pay effect is still a practically significant offsetting factor, though only significant at a 90% confidence level. Examining a particular observation can help clarify these findings. Consider Texas in 2020, wherein local truckers had 4,046 accidents. Supposing a 100-point increase in HHI, the direct association of market concentration would predict 224 fewer accidents. However, when accounting for the indirect effects via driver pay, the change in accidents associated with the same change in HHI was 201 fewer accidents. More generally, the association of higher market concentration with fewer accidents is 10.2% smaller in magnitude once the indirect effect via driver pay was considered. The magnitude of this reduction was somewhat less for injuries, at 8.8%.

The structure of these models allows for additional mediation tests. Both productivity and interstate driver pay had indirect paths to crashes and injuries via local driver pay. These paths were all significant at a 95% confidence level. Increased interstate driver pay was associated with fewer local crashes and injuries via its positive association with local driver pay. A $1,000 dollar increase in average annual pay for long-distance drivers was associated with a 0.7% reduction in crash incidents and a 0.8% reduction in injuries.

Increased productivity was associated with more crashes and injuries via its negative association with local driver pay. For productivity, the indirect and direct effects were of the same sign. Because of this, accounting for the mediating role of local driver pay resulted in a total effect larger than the direct effect alone. Including the indirect effect for the number of crash incidents resulted in a total effect that was 11.3% larger than the direct effect alone. For injuries, the increase was somewhat smaller, at 10.8%. Table 3 contains the coefficients and p-values for these associations.

Table 3. Activity level and local driver pay mediation significance tests

	Log of activity level indirect effect (via local driver pay)		Log of activity level direct effect		Log of activity level total effect	
	Coeff.	p-value	Coeff.	p-value	Coeff.	p-value
Model 1: crash incidents	0.025	0.024	0.254	0.001	0.279	0.000
Model 2: injuries	0.030	0.027	0.291	0.001	0.321	0.000

Bootstrap standard errors use 3,000 replications.

Discussion

This study melds two distinct strands of trucking safety research: research showing that bigger firms are safer, and research showing that lower driver pay increases safety risks. The results of this study confirm that these two results still stand while being estimated simultaneously. While firm size is not addressed directly, more concentrated labor markets tend to have fewer and larger firms, and the GESMs tested here show that more concentrated labor markets tend to be safer. Likewise, remuneration's relation to safety is similar to what Ju and Belzer (2024) found in the only other study of this relationship specifically for local trucking.

This study shows that local trucking labor markets that are more concentrated tend to have fewer injuries and crashes. Yet, increased market concentration also comes with lower driver pay, and lower driver pay comes with more injuries and crashes. Taken together, the balance of these contending forces still indicates that bigger is better for safety in local trucking. The indirect driver pay effect is not large enough to offset the direct effects of market concentration. Nonetheless, the indirect concentration-earnings effect reduces the safety advantages of market concentration. So, while bigger may be better for safety in local trucking, bigger would be better still if earnings were prevented from falling due to firm monopsony power.

The indirect driver pay effect presents a clear potential for increased collective bargaining in local trucking to simultaneously improve remuneration and safety. Despite a generally low degree of labor market concentration in local trucking, HII still has a measurable association with remuneration, which ought not be present under perfect competition. Labor market institutions, particularly collective bargaining, can serve to moderate employer wage-setting power in monopsonistic labor markets (Sokolova and Sorensen 2021). If unionization were to successfully moderate the driver pay effects of market concentration in local trucking, safety could be further improved.

The association of market concentration's association with driver pay may inform industry regulators. Antitrust laws apply to labor markets, but in practice, labor market cases are rare (Manning 2021). These findings suggest that antitrust regulators ought to place greater scrutiny on local trucking labor markets, even if they would be considered competitive by traditional thresholds. The models in this study still suggest that these employers have labor market power even if HHI values are low. Of course, this could be better examined with establishment-level employment data.

These results also highlight the interconnected nature of local and long-distance trucking. The labor market for local drivers and the labor market for long-distance drivers are related; the pay offered in one segment of the industry is entangled with that of the other. Still, differences in job quality or major differences in working hours seem to sustain an earnings premium for long-distance drivers. This study demonstrates that falling pay in the long-distance sector is significantly associated with more injuries and accidents in the local sector via the effect of long-distance driver pay's effect on local driver pay. Therefore, policies, or business practices in the long-distance segment of the industry are likely to have safety implications for the local segment as well.

The inverse relationship between productivity and local driver pay may seem counterintuitive. In most contexts, we would assume that increased productivity would be associated with increased earnings. Yet, there is good reason for the opposite to be the case in trucking. Drivers have target incomes they would like to achieve. With piece-rate compensation, a reduction in piece rates may lead them to work longer hours or more days or drive more intensely to hit their earnings targets. Legal restraints on driving hours may mean that drivers may still fall short of their earnings target despite these speed-ups and stretch-outs. The significance and sign on productivity suggest that the legal constraints are binding and limit drivers' ability to achieve earnings targets.

Conclusion

This study suggests that market concentration is a determinant of average driver pay and safety outcomes in local trucking. More concentrated markets tend to have fewer crashes and injuries but lower average driver pay. By way of that lower average pay, the reduction of crashes and injuries associated with increased market concentration was approximately 8% to 10% less than it would be if pay were not to fall. Fewer larger firms are likely to be better for safety, but the fall in pay that accompanies bigness may also diminish potential safety improvements. Moreover, remuneration in the long-distance sector may influence local sector remuneration, and thereby safety.

While the net effect of increased concentration is increased safety, the driver-pay effect is an important opposing factor. This result highlights the importance of considering the earnings implications of mergers and acquisitions, including safety implications, that may spill over into closely associated labor markets. These findings also underscore the importance of labor market institutions such as collective bargaining, which provide a check on employer labor market power.

Acknowledgments

Research reported in this publication was supported by the National Institute for Occupational Safety and Health (NIOSH) under Federal Training Grant T42OH008414. The content is solely the responsibility of the authors and does not necessarily represent the official views of NIOSH.

Correspondence regarding this article should be addressed to Kevin L. Conner, Rockey Mountain Center for Occupational and Environmental Health, University of Utah, 250 E. 200 S., Suite 100, Salt Lake City, UT 84111. Email: kevin.conner@utah.edu

References

Albert, Jeffrey M. 2012. "Distribution-Free Mediation Analysis for Nonlinear Models with Confounding." *Epidemiology* 23 (6): 879–88.

Ashenfelter, Orley C., Henry S. Farber, and Michael R. Ransom. 2010. "Modern Models of Monopsony in Labor Markets: A Brief Survey." *SSRN Electronic Journal.* https://docs.iza.org/dp4915.pdf

Baum, Christopher F. 2021. "Empirical Methods in Social Epidemiology." In *Handbook of Research Methods and Applications in Empirical Microeconomics*, edited by Michael A. Thornton and Nigar Hashimzade. Edward Elgar.

Belman, Dale L., and Kristen A. Monaco. 2001. "The Effects of Deregulation, de-Unionization, Technology, and Human Capital on the Work and Work Lives of Truck Drivers." *ILR Review* 54 (2A): 502–24.

Belzer, Michael H. 2023. *Sweatshops on Wheels: Winners and Losers in Trucking Deregulation*. Oxford University Press.

Belzer, Michael H., and Stanley A Sedo. 2018. "Why Do Long Distance Truck Drivers Work Extremely Long Hours?" *The Economic and Labour Relations Review* 29 (1): 59–79.

Britto, Rodrigo A., Thomas M. Corsi, and Curtis M. Grimm. 2010. "The Relationship between Motor Carrier Financial Performance and Safety Performance." *Transportation Journal* 49 (4): 42–51.

Bunn, Terry L., Svetla Slavova, and Peter J. Rock. 2019. "Association between Commercial Vehicle Driver At-Fault Crashes Involving Sleepiness/Fatigue and Proximity to Rest Areas and Truck Stops." *Accident Analysis & Prevention* 126 (May): 3–9.

Bunn, T.L., S. Slavova, T.W. Struttmann, and S.R. Browning. 2005. "Sleepiness/Fatigue and Distraction/Inattention as Factors for Fatal versus Nonfatal Commercial Motor Vehicle Driver Injuries." *Accident Analysis & Prevention* 37 (5): 862–69.

Burks, Stephen V., Michael Belzer, Quon Kwan, Stephanie Pratt, and Sandra Shackelford. 2010. *Trucking 101: An Industry Primer.* Transportation Research Circular (E-C146). Transportation Research Board.

Camerer, C., L. Babcock, G. Loewenstein, and R. Thaler. 1997. "Labor Supply of New York City Cabdrivers: One Day at a Time." *The Quarterly Journal of Economics* 112 (2): 407–41.

Cameron, A. Colin, and P. K. Trivedi. 2022. *Microeconometrics Using Stata.* 2nd ed. Vol. 2. Stata Press.

Cantor, David E., Thomas M. Corsi, Curtis M. Grimm, and Prabhjot Singh. 2016. "Technology, Firm Size, and Safety: Theory and Empirical Evidence from the US Motor-Carrier Industry." *Transportation Journal* 55 (2): 149–67.

Cinelli, Carlos, Andrew Forney, and Judea Pearl. 2022. "A Crash Course in Good and Bad Controls." *Sociological Methods & Research* 53 (3): 1071–1104.

Code of Federal Regulations. 2024. Pub. L. No. 390.5T.

Corsi, Thomas M., Curtis M. Grimm, David E. Cantor, and Dale Sienicki. 2012. "Safety Performance Differences between Unionized and Non-Union Motor Carriers." *Transportation Research Part E: Logistics and Transportation Review* 48 (4): 807–16.

Dempsey, Paul Stephen. 2003. "Transportation: A Legal History." *Transportational Law Journal* 30 (2–3): 237–366.

Department of Justice and Federal Trade Commission (DOJ-FTC). 2023. Merger Guidelines. https://www.ftc.gov/system/files/ftc_gov/pdf/2023_merger_guidelines_final_12.18.2023.pdf

Engel, Cynthia. 1998. "Competition Drives the Trucking Industry." *Monthly Labor Review* 121 (4): 34–41.

Fairchild, Amanda J, and Heather L McDaniel. 2017. "Best (but Oft-Forgotten) Practices: Mediation Analysis." *The American Journal of Clinical Nutrition* 105 (6): 1259–71.

Faulkiner, Michael R, and Michael H Belzer. 2019. "Returns to Compensation in Trucking: Does Safety Pay?" *The Economic and Labour Relations Review* 30 (2): 262–84.

Federal Highway Administration (FHA). 2022. *Highway Statistics 2022*. https://www.fhwa.dot.gov/policyinformation/statistics/2022

Federal Highway Administration (FHA). 2023. *Freight Analysis Framework*. https://ops.fhwa.dot.gov/freight/freight_analysis/faf

Federal Motor Carrier Safety Administration (FMCSA). 2019. *MCMIS Catalog and Documentation*. https://ask.fmcsa.dot.gov/app/mcmiscatalog/mcmishome

Federal Motor Carrier Safety Administration (FMCSA). 2022a. *Hours of Service*. https://www.fmcsa.dot.gov/regulations/hours-of-service

Federal Motor Carrier Safety Administration (FMCSA). 2022b. *Large Truck and Bus Crash Facts 2020*. https://www.fmcsa.dot.gov/safety/data-and-statistics/large-truck-and-bus-crash-facts-2020#A1

Jovanis, Paul P., Kun-Feng Wu, and Chen Chen. 2011. *Hours of Service and Driver Fatigue: Driver Characteristics Research*. Federal Motor Carrier Safety Administration Technical Report No. FMCSA-RRR-11-018.

Ju, Shengyang, and Michael H. Belzer. 2024. "Follow the Money: Trucker Pay Incentives, Working Time, and Safety." *The Economic and Labour Relations Review* 35 (1): 7–26.

Kudo, Takahiko, and Michael H Belzer. 2019a. "Safe Rates and Unpaid Labour: Non-Driving Pay and Truck Driver Work Hours." *The Economic and Labour Relations Review* 30 (4):

Kudo, Takahiko, and Michael H. Belzer. 2019b. "The Association between Truck Driver Compensation and Safety Performance." *Safety Science* 120 (December): 447–55.

Manning, Alan. 2003. "The Real Thin Theory: Monopsony in Modern Labour Markets." *Labour Economics* 10 (2): 105–31.

Manning, Alan. 2020. "Monopsony in Labor Markets: A Review." *ILR Review* 74 (1): 3–26.

Meiselbach, Mark K., Matthew D. Eisenberg, Ge Bai, Aditi Sen, and Gerard F. Anderson. 2021. "Labor Market Concentration and Worker Contributions to Health Insurance Premiums." *Medical Care Research and Review* 79 (2): 198–206.

Miller, Jason W., and John P. Saldanha. 2016. "A New Look at the Longitudinal Relationship between Motor Carrier Financial Performance and Safety." *Journal of Business Logistics* 37 (3): 284–306.

Muthén, Bengt, and Tihomir Asparouhov. 2014. "Causal Effects in Mediation Modeling: An Introduction with Applications to Latent Variables." *Structural Equation Modeling: A Multidisciplinary Journal* 22 (1): 12–23.

Robinson, Joan. 1969. *The Economics of Imperfect Competition.* 2nd ed. St. Martin's Press.

Rodríguez, Daniel A., Marta Rocha, Asad J. Khattak, and Michael H. Belzer. 2003. "Effects of Truck Driver Wages and Working Conditions on Highway Safety: Case Study." *Transportation Research Record: Journal of the Transportation Research Board* 1833 (1): 95–102.

Rodríguez, Daniel A., Felipe Targa, and Michael H. Belzer. 2006. "Pay Incentives and Truck Driver Safety: A Case Study." *ILR Review* 59 (2): 205–25.

Ryley, Walter T., and Michael H. Belzer. 2023. "Compensation and Crash Incidence: Evidence from the National Survey of Driver Wages." *The Economic and Labour Relations Review* 34 (1): 118–39.

Sokolova, Anna, and Todd Sorensen. 2020. "Monopsony in Labor Markets: A Meta-Analysis." *ILR Review* 74 (1): 27–55.

Soliani, Rodrigo Duarte, Luiz Silva, and Anrafel De Souza Barbosa. 2023. The Effects of Fatigue on Truck Drivers in Cargo Transportation: A Literature Review. *Interciencia* 48 (5): 228–35.

Tang, Alex P., and Yulong Ma. 2002. "The Intrastate Deregulation and the Operating Performance of Trucking Firms." *American Business Review* 20 (1): 33–42.

Torregroza-Vargas, Nathaly M., Juan Pablo Bocarejo, and Juan P. Ramos-Bonilla. 2014. "Fatigue and Crashes: The Case of Freight Transport in Colombia." *Accident Analysis & Prevention* 72 (November): 440–48.

Trick, Steven, and James Peoples. 2012. "Union Compensation Following Intrastate Deregulation: Evidence from the US Trucking Industry." *Transport Policy* 24 (November): 10–18.

US Bureau of Labor Statistics (US BLS). 2023a. *Consumer Price Indexes (CPI and PCE).* Federal Reserve Economic Data. https://fred.stlouisfed.org/categories/9

US Bureau of Labor Statistics (US BLS). 2023b. *Local Area Unemployment Statistics.* https://www.bls.gov/lau

US Bureau of Labor Statistics (US BLS). 2024. *Industries at A Glance: Truck Transportation: NAICS 484.* https://www.bls.gov/iag/tgs/iag484.htm#workforce

US Census Bureau. 2024. *County Business Patterns.* https://www.census.gov/programs-surveys/cbp.html

Viscelli, Steve. 2016. *The Big Rig: Trucking and the Decline of the American Dream.* University of California Press.

XIII. Close to the Edge: The Working Poor and the Precariousness of Employment

The Working Poor in Latin America: Causes, Factors, Dynamics, and the Provision of Income Guarantees

ELEONORA PELIZA
Pontificia Universidad Católica

Abstract

Over the past decades, there has been a global trend in the increase of poverty and different types of inequality, which has been aggravated by the pandemic. Traditionally, poverty has been associated with a lack of income due to unemployment. However, today throughout the world, and particularly in Latin America, poverty is to a large extent a problem of people who are employed, but the general concept that relates poverty to unemployment and inactivity has come under discussion, given that working is the main, if not the only, source of subsistence/income for most of the global population. In fact, labor income accounts for an average of 80% of total income in households in Latin America—it may be asserted that inequalities arise from unequal access to work and in-work poverty (i.e., poverty among people who have jobs): one-third of people who live in extreme or moderate poverty have jobs. The proliferation of different kinds of precarious jobs, lack of full and productive employment, inadequate macroeconomic and tax policies, lack of sustainable industrial policies, and a structurally ineffective transformation of economies are multifactor causes of poverty and inequality in Latin America. This paper explores the complex and dynamic causes that give rise to the working-poor phenomenon in that region. Our purpose is to analyze and understand the underlying factors that contribute to the proliferation of this problem at the regional level and propose efficient policy measures to help guarantee income in order to address this situation.

Introduction

Over the past decades, there has been a global trend in the increase of poverty and different ways of inequality, which has been aggravated by the pandemic.[1] By the end of 2022, more than 700 million workers lived in moderate poverty and 214 million were extremely poor, i.e., having an income of less than US$1.90 a day per individual in parity conditions of purchasing power.[2]

Most notably, in Latin America, 32.1% of the total region's population lives in poverty, and 13.1% in extreme poverty, which equals 201 and 86 million people, respectively. These figures are even greater if people under 18 years old are taken into consideration: 45.4% are poor and 18.5% are extremely poor.[3] Unfortunately, these data and figures are translated into a deeply disturbing reality: 21% of the Latin American population (approximately 117.3 million people) and more than 50% of the Caribbean population (13.9 million people) cannot afford a healthy diet (Food and Agriculture Organization of the United Nations).

From a dynamic perspective, work has been considered the main driver to escape poverty and social exclusion.[4] In fact, poverty has been traditionally associated with a lack of income due to unemployment. However, the general idea that relates poverty to a situation of unemployment and inactivity has been under discussion. In particular, working is the main source of subsistence/income, if not the only source for most of the global population. In fact, labor income accounts for an average of 80% of total income in the households of the region, and it may be asserted that these inequalities arise from unequal access to work and to in-work poverty (i.e., the poverty of people who have jobs).

For a significant part of the Latin American population, working for remuneration is no guarantee of a way out of poverty or extreme poverty due to the conditions under which it is performed.[5]

In this way, poverty is, in great part, a problem for working people in the region: a third of the people who live in extreme or moderate poverty have jobs. The proliferation of different kinds of precarious jobs, the lack of full and productive employment, inadequate macroeconomic and tax policies, the lack of sustainable industrial policies, and a structurally ineffective transformation of economies are multifactor causes of poverty and inequality faced by the region.

Specifically, this paper intends to explore the complex and dynamic causes that give rise to the working-poor phenomenon in Latin America.[6] Our purpose is to analyze and understand the underlying factors that contribute to the proliferation of this problem at a regional level and propose efficient policy measures to help guarantee income in order to address this problem.

Concepts and Methodology

Even though studies about inequality and poverty have always considered occupational dynamics, the concept of the working poor is relatively new. In agreement with Philippe Auvergnon,[7] we should point out that the term "working poor" is not a legal concept itself but refers to a social phenomenon characterized by people who, despite working and receiving a salary, have a lifestyle is below the poverty line. It is certainly a hybrid concept that combines the analysis of the labor market with the study of poverty. This way, we are facing a phenomenon that may be analyzed from several perspectives, given the combination of factors and elements ranging from the labor environment (low salaries, job quality, segmentation), the economy, the distribution of wealth and the impact on the development-social security, the efficiency of social protection and tax redistribution instruments and, above all, sociology, poverty, and social exclusion. The relevance of the study of this "new type of poverty" does not only consist of the quantitative increase in Occidental countries but also, and in particular, the sociological expansion to segments that have historically been part of the middle-class sector, but that, either due to new sociological realities or because of the new dynamics of labor markets, fall into poverty or are at risk of falling into poverty and social exclusion.[8]

Who is considered the working poor for this paper? The poverty phenomenon is generally addressed from a twofold analytical approach: absolute and relative. From an absolute approach, poverty is defined as the impossibility of satisfying a series of basic needs related to material subsistence (i.e., a lack of sufficient resources to live). From a relative approach, poor people are those who do not achieve an acceptable standard of living for the society in which they live. This phenomenon may also be analyzed from two points of view: a collective aspect and a family aspect, related to household income and individual income, which emphasizes the individual needs of the worker, in comparison to their income. From the collective/family perspective, a household is considered poor when its total income is insufficient to cover a basket of goods and services to satisfy the basic food and nonfood needs of a household. They are active workers who live in poor households. From this perspective, it is considered that a worker who belongs to a poor household is poor, even though their work income is sufficient to satisfy their basic needs. From the individual analysis, poor workers are those active workers whose incomes are lower than a certain line. A worker may even be considered poor if their work income is below the individual poverty line, regardless of whether they belong to a poor household. In our case, we will use the methodology used by the Economic Commission for Latin America and the Caribbean (ECLAC) and the ILO. In this way, we will take into consideration both the household characteristics of the working poor as their individual characteristics, the specific impact of poverty on different groups of workers, characterized by their personal attributes, and according to the composition of the households where they live. The working poor are all those active workers who live in poor households, even if their income is sufficiently high to cover their own basic needs.

Poverty in the Region: Factors and Dynamics

The phenomenon of the working poor[9] is not entirely new in the region. In the 1970s, the Regional Employment Program for Latin America and the Caribbean (PREALC) already raised concerns about this reality,[10] which today is even worse because of the constant increase in unemployment, underemployment, informality, the process of precarious labor conditions,[11] the consequences of the pandemic, and the impact of inflation.

As we have already explained, there are multiple factors, either social, economic, family, individual, demographic, or gender, related to working poverty. By way of example, it has been pointed out that Latin America is associated with poverty both in general terms and in terms of work, in children and teenagers less than 15 years old in their homes. The presence of minors at home reduces the possibility of escaping from poverty and increases the possibility of falling into it. There are also several studies about in-work poverty and the association with causes or dynamics that influence its shaping about ethnicity, race, age, territory, disability, immigration status, gender, sexual orientation, etc. Specifically, without disregarding the influence and incidence of other well-known factors that will be analyzed in this paper, there are factors and dynamics that are considered conclusive for the proliferation of the workers' problem in the region.

We will refer to the characteristics of the production system and productivity gaps (structural heterogeneity given by the productive and technological gaps among predominant economic units and low labor productivity); the characteristics of the labor market (heterogeneity and segmentation of the occupational structure,[12] informality, underemployment and labor precarious conditions); lack of education and basic employment qualifications, given the evident relationship between a worker's educational level and poverty; and economic aspects such as the almost nonexistent economy growth and high levels of inflation (Latin America is one of the regions that most struggles for long-term growth and that has the highest levels of inflation, which affects salary and, in particular, the food component of the market basket, as well as the most vulnerable middle-income social class).

Characteristics of the Production System

The Latin American productive structure is characterized by structural heterogeneity[13] caused by the productive and technological asymmetries among companies (i.e., the coexistence of social classes with very different productivity). Some suggest that this heterogeneity is a consequence of how the Latin American economy is inserted into the global market and its capacity to promote technical change among different sectors and production areas.[14]

The economic structure of the main Latin American countries is made up of a primitive sector, having productivity levels and income per habitant that merely allows subsistence; an intermediate sector, consisting of industries with productivity levels similar to the average productivity level of the national system; and a modern social class, including exporters, which operates with productivity levels equal to or higher than the average level of developed economies.

For instance, as opposed to what has happened in the United States over the past 30 years, Latin America has not registered a structural change in the manufacturing sector. On the contrary, industries leading the growth in the region have always been those that intensively work on natural resources, and this has caused a pure increase in productivity. Most developed countries have successfully modified their productive structure by moving toward technological sectors and changing the business model of their companies.[15] Since the 2008 crisis, the region has experienced an increase in the proportion of workers in low-productivity sectors, which has also triggered a more uneven distribution of wages.

In addition, as opposed to central economies (the United States, Canada, Japan, Germany, France, etc.) that, over recent decades have shown a trend toward homogenization, Latin American economies have enhanced heterogeneity, resulting from more and more sectors cutting off from the rest of the economy. Some authors explain that the development of the transnational (modern) sector is directly or indirectly boosted to the detriment of national sectors—middle and primitive—in the region, which gives rise to underemployment, unemployment, and an unequal distribution of income.[16]

The information on the Latin American production structure under analysis and the employment information evidence that the sectors where productivity grows account for a very scarce portion of total employment. This means that few workers are benefiting from the increase in wages, and the remaining workers are still working in sectors where productivity and wages are low.

Because most people subsist thanks to being part of the labor market, the grounds for this wrongful operation are the mechanisms that link production and employment dimensions. Productive employment has been defined as such employment yielding sufficient returns to labor to permit a worker and their dependents a level of consumption above the poverty line.[17] Labor productivity, which is generally expressed as the relation between the total value of goods and services produced in a country and labor supplies used in this production, is one of the most significant mechanisms to communicate the growth of well-being. The relation between wages and productivity is an accurate key factor in the quality of life of working people, as well as the distribution of income between work and capital, also known as remuneration, paid to factors of production.[18]

The ILO has already warned that in the region, the product per worker is similar to the figure achieved in 1980,[19] which means the possibility of progress is the same as two decades ago, explaining the reasons Latin American labor markets do not yield sufficient profits.

At a macroeconomic level, productivity is generated within the companies, and it is where we may find several of their most important key factors, in particular, those related to the labor market. For this reason, another factor that affects the working-poor phenomenon is the characteristics of the predominant economic units in our region. What does this mean? This means that companies for which the working poor work are also "poor" companies: they carry out activities with quite low labor productivity or related to family strategies of subsistence. The low productivity of informal and formal units is not merely a consequence of the workers' basic qualifications (to which we will refer below), but also that such productive units are inefficient and barely profitable, either because of a low or inexistent use of capital, inadequate and obsolete technology, difficulties in accessing supplies, and competition with formal companies. Without innovation, there is no demand for training, and without a qualified workforce, it is hard to launch innovation processes that include all workers.[20] This way, it is evident that these productive units can only generate low income for those who work there, both for owners and their dependents. In our region, a worker produces a production level of around US$10,500 per year, an amount that would be sufficient to afford an average monthly remuneration of a bit more than US$430, assuming half of the production is directed to the workforce and the other half to factors of production. Average remuneration in other developed countries or regions, such as the United States, Japan, or Europe, is much higher than in Latin America, especially because production is organized in such a manner that the outcome per worker exceeds US$40,000, and even up to US$50,000 dollars per year. For instance, in the United States, almost 70% of production is directed to compensate the workforce.

Characteristics of the Labor Market

The lack of occupational homogeneity and the twofold structure of the labor market are associated with the presence of two big sectors: primary and secondary sectors. There are different labor segments with job positions of different quality. The primary segment comprises the "good" jobs (high wages, stability, opportunity to advance, etc.), whereas the secondary segment is made up of the "bad" jobs (low wages, instability, lack of opportunities to advance, etc.).

The 2023 Labour Overview of Latin America and the Caribbean has shown that "the most urgent labour problem for the region is the employment quality and the insufficient labor and total income yielded by workers and their families." In fact, the market is one of the main factors that impact the poverty levels of a country,[21] and those factors are related to the precarious labor conditions associated with informal employment.

The United Nations ECLAC concluded that the informal sector is a result of the pressure exercised by the surplus of workforce offers and the poor creation of job positions, particularly good ones. Therefore, the heterogeneity of the economic structure is translated into a situation of labor heterogeneity.[22]

This divided structure of the Latin American labor market and the lower levels of remuneration in comparison with more developed countries have always been traditionally associated with informality and precarious labor conditions. Given the incidence of this phenomenon, many people agree that informality is now a structural feature of the Latin American economy.[23] Of Latin American workers, 51.3% work under informal conditions.[24] This way, absolute poverty and relative inequality of labor income are associated with structurally unsatisfactory labor markets,[25] so there is a large gap in the region in the income distribution, in particular, the distribution of remuneration originating from the notable differentiation of occupational structure. It has also been pointed out that labor informality is also a "phenomenon of small companies." In our region, most workers do not have social protection and have low wages, generally working for companies with fewer than five workers.

It is worth mentioning that the high levels of informality are a consequence of a great number of companies and individuals opting to be excluded from formal regimes,[26] which implies questioning the state and its capacity to enforce rules. Indeed, informality has been considered a cultural phenomenon. It has also been held that the informal sector in countries of the region is translated into exclusion and escape factors.[27] Regarding the first factor of exclusion, authors have stated that the workers or companies are pushed toward informality due to the exclusion of basic and essential social benefits. Concerning the escape factor, it is considered that the workers or companies assess the cost–benefit ratio of the convenience of being part of the formal sector or not. Other more critical authors argue that the growth of the informal sector is mainly a consequence of the inefficiencies of social policies and the expansion of noncontributing social protection programs that would work as an incentive for workers to choose informal labor, by providing social benefits without the need to contribute to the social security system.[28] In other words, business owners and workers, and citizens in general, make an "analysis" of the cost–benefit ratio as to whether it is convenient to be included in formal regimes and, based on the socio-labor reality of the region, they evidently prefer not to do it.

On top of this, two fundamental aspects arising from informality should not be disregarded. The first is that, as the formal economy evolves, the informal economy also does, and what matters is the relation to which one grows with respect to the other. The ILO Labour Overviews for the region over the past decade accurately reveal that, when the unemployment rate slows down, the creation of new job positions arises first in the informal sector rather than in the formal sector of companies.

An example of what we have been saying is what happened during the second half of 2020, in the middle of the pandemic, that informal employment was exacerbated with greater force.[29] Once the pandemic was over in 2022, the modest recovery of labor markets was driven by informal occupations, which represented around 40 to 80% of job positions created in the region.[30] The ILO Labour Overviews for the region over the past decade reveal that when the employment rate drops, the creation of new job positions arises in the informal sector instead of the formal sector of companies. In addition, the informal sector has expanded, especially among young people, given that the informal sector is the gateway to the labor market.

The second aspect is that informality is the alternative to unemployment for the poor sector,[31] the so-called informal economy or sector takes up plenty of workers that otherwise would not have the possibility of having a formal job, moving to the queue of the unemployed. In many of the countries of the region, working in the so-called informal economy is not a choice but a way of living and escaping from the calamity of unemployment. As Beccaria and Maurizio remark, in a context of insufficient demand for employment, companies from the formal sector have "factual" facilities to hire wage earners without obeying labor rules, in addition to making working conditions even worse for job positions, which has direct consequences in the income and segmentation of the labor market. Informal wage earners represent 30% of the total urban employment of the region and half of the informal employment, ranging from 17% (Chile) to more than 40% (Bolivia, Ecuador, Guatemala, Mexico, Nicaragua, Paraguay, and Peru).

In the region, there are more and more occupations, somehow associated with medium- and large-sized companies of the formal sector that, even though they cannot be considered strictly informal, do not have precarious conditions features or signs such as a certain period or part-time working schedules, zero-hour contracts, or job positions related to platform economies. Even though these new forms of working are still limited in the region, they all have a degree of insecure income or remunerations lower than other workers with similar qualifications, which makes them likelier to fall into poverty or low-income situations.[32] Expressed differently, these new forms or systems of employment that cannot be standardized (economies of temporary occupations) cannot be differentiated from predominant precarious and informal forms of employment.

The other group, comprising 53.1% of informal employment, is made up of independent informal workers, which are calculated at around an average 23.1% of urban employment. Such numbers vary with figures ranging from less than 22% (in Argentina, Brazil, Chile, and Uruguay) to more than 35% (in Bolivia, Colombia, Peru, Dominican Republic, and Venezuela). Since the beginning of the 21st century, there has been a gradual increase in self-employment, which was higher than in most of the countries of our region. The greatest numbers have been in Argentina, Colombia, Costa Rica, Peru, and Venezuela.

Salary Penalty, Informality, Underemployment, and Poverty

The relation between precarious jobs, informality, and poverty is a longtime concept: 80% of poor workers work under informal conditions.[33] The regional director of the ILO has also remarked that, according to the ILO overviews, "informal workers have around three and four times more likelihood of being poor than formal workers, and also explain that around 70 and 90 per cent of total employment poverty."[34] Informality entails a "salary penalty" that is reflected in lower labor income per hour for workers under these conditions, which ranges between 20 and 35% in countries with available data. This means that the less labor income an informal worker earns with respect to other workers carrying out the same duties in the formal sector, how much they stop earning is the amount this worker loses for being an informal worker in comparison to its formal counterpart.

Salary Penalty Associated with Informal labor

If we consider only the aggregate informal workers, for instance, in Argentina, workers of the informal sector represent monthly incomes 65% lower than the ones received by employed people of the formal sector in the whole country.[35] During the pandemic, all countries of the region were affected by labor poverty, even though such countries with the highest levels of informality were greater affected, such as Colombia and Peru, which reached labor poverty levels of more than 60% and 40%,[36] respectively, at the worst moment of the pandemic.

As pointed out by Poy, the heterogeneity of the economic-occupational structure reproduces a harsh pattern of socio-labor inequality, which has direct consequences on the family's capacity for subsistence.[37] Informal nonwage earners represent most of the active workers, followed by informal wage earners. Informal nonwage earners represent the greatest part of the poor, active workers, varying depending on the country. For instance, in Argentina, two of every five poor workers are informal nonwage earners. In Brazil, this figure represents half of them, and in Paraguay, three of every four poor workers are informal, nonwage earners.

Regarding informal wage earners, the proportion of labor poverty is around 22% in Brazil, 28% in Argentina, and 19% in Paraguay. Considering wage earners and informal nonwage earners, they represent 73% of labor poverty in Brazil, 68% in Argentina, and 94% in Paraguay. Finally, we cannot ignore the fact that there is a portion of formal workers living in poor households. In Argentina, 32% of poor workers are formal workers; in Brazil, 27%; and in Paraguay around 6%.

Underemployment

The last special labor overview published by the ILO reveals that the impact of poverty is even greater for part-time workers in relation to those who work full-time or who work more than 45 hours per week. In Argentina and Brazil, the probability of living in poverty is even higher when part-time employment is involuntary. The lack of sufficiently long working schedules arising from low wages has negative effects on the workers' capacity to gain enough income so as not to live in poverty.

Everything that has been explained and described shows that inequality in income cannot be explained without understanding the quality and productivity of job positions among the different productive industries, which is projected in also very unequal earnings among workers, capital and work, and the heterogeneity of occupational and productive structure causes more poverty. It has been explained that the persistence of occupations linked to the micro-informal sector and the insufficient demand for employment of quality has caused "selective processes of poverty" that would affect households having a workforce assigned to such job positions.[38]

Basic Employment Qualifications

Latin America is the region of the world with the highest educational gaps,[39] one of them is the quality and relevance of education, which causes high levels of income inequality.[40] Education and professional training are conclusive factors for labor income and its unequal distribution. There is a correlation between the educational levels of workers and poverty. Even though there have been some advances over the past 20 years, especially in access to primary and secondary school education, there are still many challenges pending.

These inequalities are linked to disparities in the educational system, particularly in the following areas:

- Content and quality: There is a tendency to focus solely on basic skills, such as literacy and mathematics
- Infrastructure: Differences in school size and building conditions.
- Faculty education: Variations in teachers' qualifications and training.
- Student learning outcomes: Differences in academic achievement.

These disparities are reflected in the results of the Programme for International Student Assessment (PISA), conducted by the Organization for Economic Co-operation and Development (OECD). On average, students in Asian countries score more than 20% higher than those in Latin America and the Caribbean.[41]

In this way, the level of Latin American workforce skills is still low in comparison to developed countries. In most of the countries in the region, the population with basic skills is still predominant in the adult population. Only Argentina, Chile, and Panama have levels similar to the average of the OECD. The low level of skills and capabilities entails problems of employability and productivity: Companies require skills and capabilities that human resources do not have. Productivity in Latin America is 0.76% of the average total productivity of the United States. This means that using the same amount of resources (workforce and capital), the production of our region would be approximately 25% lower than in the United States.[42] It is a vicious circle; workers with limited training are those who are mostly employed in low productivity facilities. These characteristics are also present in those who work in the formal sector but under precarious conditions. Ultimately, the greatest inequality in the developing world would arise from the greatest differentiation in job offers in terms of productivity, which is also based exclusively on human resources.

Poverty associated with such low remuneration is not only the result of informality and precarious labor conditions but also of low levels of workers' qualification in the region. By way of example, we can quote the cases of Argentina, Brazil, and Paraguay. In 2021, the percentage of poor workers with basic education was around 30.5% in Argentina, 21.3% in Brazil, and 29.3% in Paraguay, whereas the impact of poverty on workers with complete higher education was 8.4% in Argentina, 2% in Brazil, and 4.3% in Paraguay.[43]

The Economy in the Long Term

Even though Latin America is in the process of recovering its economy after the pandemic, it is one of the world's regions that is having more difficulties growing in the long term. The ECLAC has estimated Latin America's economic growth at a rate of 2.7% per year,[44] which means 16 countries of the region—that is to say, half of the countries, by the end of 2022, had not recovered the GDP level they had before the pandemic. Figures for the last two quarters of 2023 confirmed the economy slowing down, which evidences that the regional GPD has been at a standstill for more than five quarters. By 2023, an average GDP of only 1.7% is expected, and by 2024, there is a slight increase in the growth rate of 1.5%.

It has also been said that "Latin America and the Caribbean's low growth may be aggravated by the negative effects of an intensification of climate shocks, if the investments that countries need in climate change adaptation and mitigation are not made."[45] If this slow level of growth is maintained, the regional economy would take more than 30 years to double its size, which means that the Latin Americans' standard of living (product per habitant) could just double in more than 100 years at this rate.[46]

As a consequence, this low growth of economic activity entails a slowdown of employment growth, estimated at 1.9% in 2023 and 1.1% in 2024. With respect to employment quality, workers are hired in less productive industries that are related to poverty and social vulnerability, which in turn has a direct impact on income. In 2019, the percentage of people living on less than US$5 per day was 28%, and 16% on US$3.1 per day.

As we have pointed out, the pandemic aggravated the tendencies toward this worsening: in 2020, in spite of the policies taken to hold back the pandemic effects, social indicators were also aggravated, 34% of Latin Americans lived on less than US$5 per day, and the figure rose to 20% for those who earned less than US$3.1 per day. The Gini coefficient reflects the highest rating of the decade: 0.52 in 2020. In 2023, around 18% of the population lives on less than US$3.1 per day. As we have stated, the poverty levels of the region have exceeded the levels from one decade ago.[47]

It is also observed that, in many Latin American countries, economic growth does not necessarily encourage the creation of new quality jobs and, consequently, does not allow for improving people's well-being. This phenomenon has been called the paradox of growth without well-being (Easterling paradox).[48] Above all, at the beginning of this century, there has been macroeconomic growth in most Latin American economies that was not translated into greater well-being for their populations, which proves that something is not working. The ILO and ECLAC have already affirmed that regional poverty is mostly linked to the trajectory of the outcome per worker rather than the production itself. In other words, the Easterling paradox observed in the region over the past few decades is related to the fact that the outcome per worker has not sufficiently recovered after the drop experienced over the 1980s. It should be understood that economic growth (population growth) does not necessarily reduce poverty. Outcome growth is merely a required condition to reduce poverty. The evolution of the outcome per worker is most closely related to poverty.

Inflation

Inflation is a regressive tax that undoubtedly has a negative effect on poverty and the operation of the labor market. A consistent increase in prices, mainly food and energy, gives room to more poverty, given that inflation reduces available income and real wages when nominal wages grow at a slower rate than prices. When inflation increases, even though nominal income remains steady, the purchasing power for goods and services is shrunk.

Since 2019, it has been seen that there has been more and more poverty, and that increase is strictly associated with price rises in the consumer price and energy price indexes (the food inflation rate has increased 30% from February 2020 and the consumer price index increased 18%, so the poorest households were the ones mostly affected because they depend on a greater extent of labor income, which were deeply harmed by inflation).

Also, since the second quarter of 2020, due to several causes (the pandemic, the conflict between the Russian Federation and Ukraine, the monetary issuance to finance different aid programs, tax incentives, and other external factors), inflation has gone up in all countries of the region, and it reached its highest peak since the 2008 financial crisis. The annual inflation rate for the average country of the region reached 9.6% in July 2022, in comparison to 7.6% (values that have not been registered since 2002).

Likewise, income may decrease during an inflation crisis as a response to adjustments in the labor market. When there is less real income in households, people must work more hours to maintain the same consumption level. This is how the inflation level has increased labor and household poverty levels. Workers are working longer and longer work shifts to maintain their consumption level.

Along this line of reasoning and the Inter-American Development Bank estimates, inflation rates would trigger an increase in poverty by 2.4% and extreme poverty by 2.5% by 2023, which means taking almost 13.8 million and 14 million people to poverty and extreme poverty.

Even though it is extremely necessary to restrain inflation in the region, it also means taking risks, as certain tightening monetary policies to keep inflation steady may negatively affect labor markets in the short term because the economy is slowed down as an inevitable result of reducing expenses, which slow down the economy and probably also generate more unemployment. Recent factual evidence shows that income decreases for all workers, given that in countries with a high informality level, a monetary tightening policy reduces income from informal workers to a greater extent than formal workers.

At the same time, it has been warned that monetary shock generates a change from formality to informality. Even though income decreases for all formal and informal workers, workers placed in the lowest segment of informality suffer relatively more than their formal counterparts.[49]

Challenges and Proposals

Over the past few decades, there have been different attempts in the region to revert these phenomena and dynamics to speed up growth and reduce poverty. That being said, it should be considered that it is not possible to reduce poverty in the long term without decent work. To reduce the working-poor phenomenon in the region, it is crucial to fight against informality, promote labor productivity, secure financial support and public investment, access to education and training for the development of new capabilities and skills, and encourage professional technical education.

Combating Informality and Promoting Productivity

To eradicate the dynamics giving rise to the working-poor phenomenon in the region, it is extremely urgent to revert to informal labor, which comprises old and new types of employment related to the impact of technology that, as we have seen, are crucial elements of the matrix of low income and social inequality. There are simply no magic tricks to reduce informal labor; comprehensive and tailored strategies are needed, which have to be designed in accordance with companies' productivity levels, workers' qualifications, tax conditions, and preexisting rules.[50] The formalization process takes time and requires multiple and several interventions in several areas.[51]

It is essential to create adequate conditions and reinforce people's capacity to afford a standard of living through quality jobs, and for that purpose, a global approach is necessary to put an end to the expansion of informal labor. Discussions should also involve authorities, companies, and unions to design regimes that promote formal employment, encourage incorporation into the formal system, and foster decent work as a mechanism to overcome poverty.

ILO Recommendation 204 Concerning Transition from the Informal to the Formal Economy[52] recognizes that "most people enter the informal economy not by choice." It also establishes certain working rules for governments, called the 12 Guiding Principles, for different policies that may be addressed for a transition from informality to formality. It also points out that the transition to formality requires designing coherent, integrated, and tailored strategies, highlighting that the key elements for a transition should include macroeconomic policies that promote employment, enhance productivity and facilitate structural transformation processes; trade, industrial, tax, sector, and infrastructure policies that promote employment; policies and institutes for the labor market; education and skills development policies and measures to facilitate the school-to-work transition and from unemployment or inactivity to work. These policies are aligned with the eighth sustainable development goal to promote sustained, inclusive, and sustainable economic growth; full, productive employment; and decent work for all. Another of the challenges to face the causes and factors that give rise to the working-poor phenomenon is to modify the characteristics of the productive matrix. The low productivity of a significant part of economic units hampers the generation of sufficient aggregate value for the development of formal employment. In addition, the expansion of informal labor in economic units of low-productivity facilities that highly productive companies hire workers under informal conditions.[53]

In fact, the existence and persistence of productivity gaps in almost the entire productive system of the region explain the external restraint to economic growth and the relative lack of strategic sectors of highly complex technology capable of more quickly disseminating the technical change to the entire productive system. To address this reality, it is essential to design productive development policies that are focused on enhancing innovation and developing new companies in key industries, where it is essential to give more importance to knowledge-intensive sectors.[54]

Following Bourguignon, we note that the changes in the poverty rate are not only associated with changes in the average income, but also the transformation of its distribution, so it is necessary to combine growth and distribution policies because the composition of the productive system is not neutral for the development process.

As has already been suggested,[55] the industrial policy should combine horizontal and sectorial actions, the first with the purpose of raising productivity levels of the entire system, facilitating access to long-term financing; promoting investment in research; fostering development and innovation; granting subsidies to workforce education and training. The second should focus on the development of strategic activities capable of increasing the generation of currencies and promoting a productive framework with the greatest intersector chains. By way of example, we note some initiatives carried out in Brazil, Ecuador, Mexico, and Peru that developed value chains linking informal sectors of the productive chains to formal ones, subjecting public acquisitions to the formalization of jobs by supplier companies. At a microeconomic level, Mexico and Brazil have emphasized business development services, as well as the financing and dissemination of technologies aimed at improving management and increasing company productivity and competitiveness.

Financial Support and Public Investment

For all that, it is necessary to have financial support and public investment. It has been said that "there is no industrial policy without a credit policy," given that supporting financial instruments are a necessary pillar for technological and productive development. It is necessary to promote investment in infrastructure through public investment, given that the evidence shows that when investment in infrastructure is backed up with mechanisms to increase efficiency, there is a positive impact on the poorest' income.[56] Financial support should also be the means to foster policies that promote bonding and transferring technology between small and medium-sized companies, universities, and institutes and the creation of education and training institutes to focus on productivity, quality, and innovation.

Access to Social Protection and Security

The extension of access to and protection for social security are considered key factors for many countries in the region to reduce labor poverty. In effect, to revert the working-poor phenomenon, strong social protection systems are needed that are capable of giving a response based on solidarity principles that help satisfy people's needs over their lifetime: through a basic level of protection[57] that is supplemented by contributing social insurance programs.

It is also necessary to adopt protection systems for new labor models, which imply more precarious conditions and income, such as atypical employment, platform workers and self-employed workers.

Over the past two decades, countries of Latin America and the Caribbean have launched several social programs aimed at overcoming poverty, increasing the number of noncontributing social protection programs, among which we can refer to programs of transfers subject to income, labor, and productive inclusion programs and social pensions. We can also outline policies carried out in Argentina, Uruguay, Colombia (special taxpayer regimes) and Brazil (MEI program) that were intended to expand the scope of working people who contribute to social security. These mechanisms and policies are directed toward active workers in micro-companies (self-employed workers, regular, or domestic service workers).

However, as has already been stated, a case-by-case analysis should be made considering the actual effectiveness. Some authors have stated that social policy may unintentionally generate incentives for informality and, consequently, aggravate labor poverty. This may be the case of conditional and nonconditional transfers for people under poverty and vulnerability conditions, given that many laws provide that monetary allowances are lost once a person gets a formal job providing sufficient income.

The negative effect in the case of monetary transfers has also been remarked, for instance, in Argentina, Colombia, and Uruguay.[58] It has been said that it is a good policy to develop assistance programs for formal employment that reward those who are under formal systems, in comparison to many current social assistance programs that generally withdraw monetary transfers once the person gets a formal job.

Access to Education

As already stated, there is a high poverty rate in the region associated with low levels of workers' qualifications. In fact, there is an evident link between schooling levels and poverty rates: the lowest educational level, the greatest tendency toward labor poverty. For this reason, it is vital to secure education and professional technical training policies. Policies shall also guarantee that there is a public and quality education offer, which is intended to provide lifelong learning, with more flexible educational systems, and with "entrances" and "exits" according to different needs. Therefore, one of the main critical aspects is to identify and characterize skills that have to promote educational programs and training programs in our region to overcome this reality.

To answer these questions, Latin American states have been questioning their educational and teaching guidelines in sociopolitical, institutional and pedagogical contexts. Skill-based teaching, which is pedagogical by nature, should be arranged with a curricular organization based on institutional skills and an educational policy focused on the sociopolitical skills model.[59]

For this purpose, it is necessary to make solid public-private alliances between the educational and productive sectors to project the evolution of the labor demand in the long term and, accordingly, the answers of training and educational systems to these future perspectives. A positive example was the enrollment expansion in public universities and high school professional training entities from 2002 through 2016 in Brazil that allowed public institutes of professional technical education to increase from 140 to 600, doubling enrollment rates from 558,000 in 2002 to more than a million in 2016.[60] In Brazil, we can point to the National Service of Industrial Education (SENAI), which is the biggest professional and technological education institute in Latin America. SENAI convenes more than 2.3 million Brazilian workers per year. It is a state-owned entity that provides support to 28 industrial areas, and the main program beneficiaries are workers, young people, and adults from the industry and trade.[61] This type of training has allowed us to address vulnerabilities, such as reconversion (*upskilling and reskilling*) in a context of so much automatization of occupations and labor insertion of elderly people (generally older than 45 years old).[62] The teaching of skills, in particular, is key to achieving workers of better quality, and for new generations to acquire the skills and capabilities that will be required by the labor market in the future.

Conclusions

As evidenced, some different factors and dynamics cause and impact the working-poor phenomenon. Even though the factors remarked on in this paper are mere indicators, we can have an overview of the noticeable vulnerability of Latin American workers. Labor poverty is translated into underdevelopment, and it is a symptom of inefficient labor and management policies.

It also reflects the inefficient state of a well-being manager. The creation of decent work is still a great structural challenge for Latin America and the Caribbean. To reverse this phenomenon, it should be taken into consideration that poverty cannot be reduced permanently without decent work. The 2020 Sustainable Development Agenda expresses an agreement on the need to foster more inclusive societies, supportive of securing to "leave no one behind" in the path toward development.

It has been said that 10,000 billion dollars are needed to eradicate extreme and moderate poverty in the region by 2030. However, it would be naïve to consider that this goal may only be achieved with income transfers.[63]

The working-poor phenomenon may not be battled if the economy does not generate room for investment and business development, so it is essential to create and strengthen competitive small- and medium-sized companies given their capacity to rapidly generate jobs.[64] Policies to improve the operation of labor markets are also needed. The solution to this problem requires much more than available resources, social security policies, and income substitution.

As we have already said, real social justice requires time and structural changes in society, so it is pivotal to apply a global approach, always considering that employment is the master key toward equality.

During his 1936 acceptance speech for presidential renomination, Franklin Roosevelt said: "Liberty requires opportunity to make a living—a living decent according to the standard of the time, a living which gives man not only enough to live by, but something to live for. True individual freedom cannot exist without economic security and independence. Necessitous men are not free men." Those words were as transcendental then as they are today.

Endnotes

[1] Vertical inequalities—among the rich and the poor—and horizontal inequalities among different groups of society.

[2] ILO. 2023. "World Employment and Social Outlook, Trends 2023." https://www.ilo.org/publications/flagship-reports/world-employment-and-social-outlook-trends-2023

[3] ILO. 2023. "World Employment and Social Outlook, Trends 2023." https://www.ilo.org/publications/flagship-reports/world-employment-and-social-outlook-trends-2023

[4] In addition to being the main driver for poverty eradication and for the creation, exacerbation, or mitigation of inequality, work is a fundamental mechanism for building autonomy and identity, upholding dignity, and expanding the scope of citizen action; it is also the main avenue for social and economic integration.

[5] L. Abramo, S. Cecchini, and B. Morales. 2019. "Social Programmes, Poverty Eradication and Labour Inclusion: Lessons from Latin America and the Caribbean." ECLAC.

[6] For this purpose, we analyzed the data obtained in the ILO Labour Estimates for the region and ECLAC (ten countries of the region: Argentina, Bolivia, Brazil, Colombia, Costa Rica, Ecuador, Paraguay, Peru, Dominican Republic and Uruguay).

[7] P. Auvergnon. 2012. "El Fenómeno de los Trabajadores Pobres: Revelador de las Funciones y Tendencias del Derecho Social." *Revista Latinoamericana de Derecho Social*, no. 14: 43–78. https://www.scielo.org.mx/scielo.php?pid=S1870-46702012000100043&script=sci_abstract&tlng=en

[8] F.J. Calvo Gallego. 2016. "Trabajadores Pobres y Pobreza De Los Ocupados: Concepto y Evolución de la Pobreza en la Ocupación: El Impacto de las Últimas Reformas Legales." *Legales Temas Laborales,* No. 134: 63–106.

[9] Working poor in the United States; precarious, vulnerable, atypical, low salaries in France and Germany.

[10] PREALC. 1974. "La Política de Empleo en América Latina: Lecciones de la Experiencia de PREALC." *El Trimestre Económico* 41, no. 164(4): 917–36.

[11] L. Muñiz Terra. 2009. "Nuevos y Viejos Escenarios en el Mundo Laboral Latinoamericano el Mundo del Trabajo en América Latina: Trabajo, Empleo, Calificaciones Profesionales, Relaciones de Trabajo e Identidades Laborales." Latin American Council of Social Sciences.

[12] The hypothesis of labor market segmentation has been analyzed for several years, and there is some research in this regard (M. Reich. 2009. *Segmented Labor Markets and Labor Mobility*. Edward Elgar).

[13] A heterogeneous productive structure exists when there are sectors, areas, or activities in which productivity is high.

[14] CAICYT- CONICET. 2017. "Trabajo y Sociedad. Sociología del Trabajo—Estudios Culturales—Narrativas Sociológicas y Literarias." No. 29.

[15] ECLAC Collection. 2010. "Heterogeneidad Estructural y Brechas de Productividad: De la Fragmenación a la Convergencia." https://archivos.juridicas.unam.mx/www/bjv/libros/12/5935/5.pdf

[16] P.I. Chena. 2016. "Heterogeneidad Estructural y Distribución del Ingreso. El Pasaje del Estructuralismo Latinoamericano al Neoestructuralismo." *En Memoria Académica*, No. 35: 6–30.

[17] ILO. No date."What Is Decent Work?" https://www.ilo.org/global/topics/decent-work/lang--es/index.htm

[18] M. González, E.A. López Cabrera, J. Antonio, and R. Cabral Torres. 2022. "Relación Entre Productividad Laboral y Remuneraciones: Un Análisis de Proximidad Espacial a Nivel Estatal en la Industria Manufacturera en México, 2004, 2009, 2014 y 2019." ECLAC. https://www.cepal.org/es/publicaciones/47894-relacion-productividad-laboral-remuneraciones-un-analisis-proximidad-espacial

[19] It is usually assumed that productivity analysis is interested only in production agents. However, there is no international evidence about the conclusive role played by labor productivity in the creation of prosperity and development, in addition to its inverse relation with poverty in countries, measured through the most common approach (the real product index per worker).

[20] ILO. 2004. "Labour Overview." https:/www.ilo.org/publications/2004-labour-overview%20

[21] L. Beccaria, R. Maurizio, A.L. Fernández, A.L. Monsalvo, and M. Álvarez. 2011."Dynamics of Poverty, Labor Market and Public Policies in Latin America." PMMA Working Paper 2011-0. Poverty and Economic Policy Research Network.

[22] ILO. 2023. "World Employment and Social Outlook: Trends 2023." https://www.ilo.org/publications/flagship-reports/world-employment-and-social-outlook-trends-2023

[23] F. Bertranou, L. Casanova, M. Jiménez, and M. Jiménez. 2013. "Informalidad, Calidad del Empleo y Segmentación Laboral en Argentina." Organización Internacional del Trabajo. https://www.ilo.org/sites/default/files/wcmsp5/groups/public/@americas/@ro-lima/@ilo-buenos_aires/documents/publication/wcms_228741.pdf

[24] A. Espejo. 2022. "Informalidad Laboral en América Latina: Propuesta Metodológica para Su Identificación a Nivel Subnacional." Documentos de Proyectos (LC/TS.2022/6). ECLAC. https://www.cepal.org/es/publicaciones/47726-informalidad-laboral-america-latina-propuesta-metodologica-su-identificacion

[25] ILO. 2017. "Perspectivas Sociales y del Empleo en el Mundo 2016: Transformar el Empleo para Erradicar la Pobreza." https://www.ilo.org/es/publications/perspectivas-sociales-y-del-empleo-en-el-mundo-2016-transformar-el-empleo

[26] A.O. Hirschman. 1970. *Exit, Voice and Loyalty: Responses to Decline in Firms, Organizations and States*. Harvard University Press.

[27] G.E. Perry, W.F. Maloney, O.S. Arias, P. Fajnzulber, A.D. Mason, and J. Saavedra-Chanduvi. 2007. "Informalidad: Escape y Exclusión." World Bank. https://documents1.worldbank.org/curated/zh/889371468313790669/pdf/400080PUB0SPAN101OFFICIAL0USE0ONLY1.pdf

[28] R. Maurizio. 2021. "Employment and Informality in Latin America and the Caribbean: An Insufficient and Unequal Recovery." Technical Note. ILO. https://www.ilo.org/sites/default/files/wcmsp5/groups/public/@americas/@ro-lima/@sro-port_of_spain/documents/genericdocument/wcms_819029.pdf

[29] ILO. 2022. "2022: Labour Overview: Latin America and the Caribbean." https://www.ilo.org/sites/default/files/wcmsp5/groups/public/@americas/@ro-lima/documents/publication/wcms_876565.pdf

[30] Organización Internacional del Trabajo. 2019. "La Informalidad en la Visión de la OIT."

[32] L. Beccaria and R. Maurizio. 2020. "Poverty, Inequality and the 'Working-Poor' Phenomenon." In *Reflections on Work: Perspectives from the Southern Cone of Latin America on the Occasion of the ILO Centenary*, F. Bertranou and A. Marinakis, eds. ILO. https://www.researchgate.net/profile/Fabio-

Bertranou/publication/340807016_Reflections_on_work_Perspectives_from_the_Southern_Cone_of_Latin_America_on_the_occasion_of_the_ILO_Centenary/links/5e9e6fdaa6fdcca7892bdf2a/Reflections-on-work-Perspectives-from-the-Southern-Cone-of-Latin-America-on-the-occasion-of-the-ILO-Centenary.pdf

[33] R. Maurizio. 2022. "Jornadas de Análisis de Mercado de Trabajo, Políticas de Formalización y Transición Justa." Organización Internacional del Trabajo. https://www.ilo.org/es/meetings-and-events/jornadas-de-analisis-de-mercado-de-trabajo-politicas-de-formalizacion-y

[34] ILO. 2023. "Informalidad y Pobreza Laboral Lastran a Mercados de Trabajo en América Latina y el Caribe." https://www.ilo.org/es/resource/news/informalidad-y-pobreza-laboral-lastran-mercados-de-trabajo-en-america

[35] M. Agustín and A.O. García. 2013. "Informalidad Laboral, Probeza y Regions: Un Análysis desde la Coyuntura Argentina." *Memoria Académica Compartimos lo que Sobemos.* No. 9: 107–25. https://www.memoria.fahce.unlp.edu.ar/art_revistas/pr.6298/pr.6298.pdf

[36] Inter-American Development Bank. 2023. "Preparing the Macroeconomic Terrain for Renewed Growth: 2023 Latin American and Caribbean Macroeconomic Report." https://flagships.iadb.org/en/MacroReport2023/preparing-the-macroeconomic-terrain-for-renewed-growth

[37] Poy, S. 2020. "Heterogeneidad Laboral y Procesos de Empobrecimiento de los Hogares en Argentina Problemas del Desarrollo." *Revista Latinoamericana de Economía* 51, no. 201.

[38] Poy, S. 2020. "Heterogeneidad Laboral y Procesos de Empobrecimiento."

[39] UNESCO-ECLAC. 2021. "Education in the Time of COVID-19 and Beyond." Policy Brief. https://www.un.org/development/desa/dspd/wp-content/uploads/sites/22/2020/08/sg_policy_brief_covid-19_and_education_august_2020.pdf

[40] ILO/Organazación Internacional del Trabajo. 2017. "The Future of Vocational Training in Latin America and the Caribbean: Overview and Strengthening Guidelines." https://www.oitcinterfor.org/en/publications/future_vt

[41] Global Education Monitoring Report Team, Laboratory of Education Research and Innovation for Latin America and the Caribbean, UNESCO Office Santiago, and Regional Bureau for Education in Latin America and the Caribbean. 2020. "Informe de Seguimiento de la Educación en el Mundo, 2020, América Latina y el Caribe: Inclusión y Educación: Todos y Todas sin Excepción." https://unesdoc.unesco.org/ark:/48223/pf0000374615

[42] G. Crespi, E. Fernández-Arias, and E. Stein, eds. 2014. *Rethinking Productive Development: Sound Policies and Institutions for Economic Transformation.* Inter-American Development Bank.

[43] ILO. 2022. "2022: Labour Overview: Latin America and the Caribbean." https://www.ilo.org/sites/default/files/wcmsp5/groups/public/@americas/@ro-lima/documents/publication/wcms_876565.pdf

[44] ECLAC. 2022. "Economic Survey of Latin America and the Caribbean 2022: Trends and Challenges of Investing for a Sustainable And Inclusive Recovery". https://www.cepal.org/en/publications/48078-economic-survey-latin-america-and-caribbean-2022-trends-and-challenges-investing

[45] ECLAC. 2023. "Economías de América Latina y el Caribe Mantendrán Bajos Niveles de Crecimiento en 2023 y 2024". https://www.cepal.org/es/comunicados/economias-america-latina-caribe-mantendran-bajos-niveles-crecimiento-2023-2024

[46] That would be the case if the population growth rate was low, which did not happen because of the population boom experienced in the region.

[47] A. Galindo and V. Nuguer 2023. "2023 Latin America and the Caribbean Macroeconomic Report." Inter-American Development Bank. https://publications.iadb.org/en/2023-latin-american-and-caribbean-macroeconomic-report-preparing-macroeconomic-terrain-renewed

[48] ILO. 2004. "2004 Labour Overview: Latin American and the Caribbean." https://www.ilo.org/publications/2004-labour-overview

[49] D.P.B. Gomes, F. Iachan, A.P. Ruhe, and C. Santos. 2023. "Monetary Policy and Labor Markets in a Developing Economy." Inter-American Development Bank. https://papers.ssrn.com/sol3/papers.cfm?abstract_id=4432540

[50] M. Cárdenas, C. Fernández, A. Raster, and D. Zamora. 2022. "Consideraciones para el Diseño de Políticas Fiscales para Reducir la Informalidad en América Latina y el Caribe." Inter-American Development Bank. https://publications.iadb.org/es/consideraciones-para-el-diseno-de-politicas-fiscales-para-reducir-la-informalidad-en-america-latina

[51] J.M. Salazar-Xirinachs and J. Chacaltana. 2018. "Informality in Latin America and the Caribbean: Why Does It Persist and How Can It Be Addressed?" ILO Regional Office for Latin America and the Caribbean. https://www.researchgate.net/publication/333681245_Informality_in_Latin_America_and_the_Caribbean_Why_does_it_persist_and_how_can_it_be_addressed

[52] Conferencia Internactional del Trabajo. No date. "Recomendácion sobre la Transición de la Economía Informal a la Economía Formal." Recomendación 204. https://www.ilo.org/wcmsp5/groups/public/---ed_norm/---relconf/documents/meetingdocument/wcms_379098.pdf

[53] Organización Internacional del Trabajo. 2022. "Sesión Especial. Productividad y Formalización Laboral." https://www.ilo.org/es/resource/sesion-especial-productividad-y-formalizacion-laboral

[54] ILO. 2022. "Sesión Especial. Productividad y Formalización Laboral." https://www.ilo.org/buenosaires/eventos-y-reuniones/WCMS_861136/lang--es/index.htm

[55] ECLAC. 2021. "El Futuro de la Matriz Productiva Argentina: Consensos y Nuevas Perspectivas." https://www.argentina.gob.ar/sites/default/files/futuros_de_la_matriz_productiva_editado.pdf

[56] c. powell ECLAC. 2021. "El Futuro de la Matriz Productiva Argentina."

[57] ILO. 2019. "Work for a Brighter Future: Global Commission on the Future of Work." https://digitallibrary.un.org/record/3827525?ln=en&v=pdf

[58] M. Cárdenas, C. Fernández, A. Rastel, and D. Zamora. 2022. "Consideraciones Para El Diseño De Políticas Fiscales." The authors point out that the universal allowance per child (Argentina), which provides monetary transfers to households with minor children, reduced formalization among those who were part of the program in comparison to those who did not. In Uruguay, a program of conditional transfers led to reductions in the formal workforce participation among all program beneficiaries and among single mothers. They also remark that there are examples of negative effects related to noncontributing benefits in Colombia and Chile. The subsidized health regime in Colombia raised the probability of informality. Something similar happened in Chile with the 2008 social security reform, which improved the pension of people who have not saved enough. The reform reduced participation of workers older than 40 years in the formal labor market by 4.1% and the probability of becoming part of the formal regime of men and women older than 56 and 65.

[59] Organización de Estados Iberoamericanos. 2020. "Miradas sobre la Educación en Iberoaméfia 2020." https://oei.int/oficinas/secretaria-general/publicaciones/informe-miradas-2020-2

[60] Mercadante. 2019. "Educaçâo e Capacitaçâo Técnica e Profissional no Brasil," Documentos de Projetos (LC/TS.2019/44). Santiago, ECLAC.

[61] For the modernization of technical assistance and education in the industry, SENAI hired the Fraunhofer IPK institute from Berlin, which transferred knowledge about teaching in planification and the creation of innovation centers. There are more than 130 projects contracted for the education of new skills and capabilities, which led to the restructuring of 28 training institutes. Some companies in the automobile industry of Brazil have been partners of SENAI from the beginning to solve the demand for new skills by agreeing on learning rules. (S. Peliza. 2023. "Formación Profesional y Educación para Nuevas Competencias como Política Pública en América Latina").

[62] S. Smeck, M. Oviedo, and A. Fiszbein. 2020. "Dual Education in Latin America: Challenges and Opportunities. Inter-American Dialogue." The Dailogue: Leadership for the Americas." https://thedialogue.org/wp-content/uploads/2019/12/Dual-education-12.9.2019-ENG.pdf

[63] United Nations. 2023. "Objtivo 1: Poner Fin a la Pobreza en Todas Sus Formas en Todo el Mundo." https://www.un.org/sustainabledevelopment/es/poverty/

[64] School of Economics, Universidad de Buenos Aires. 2016. "Open Department Plan Fénix." https://vocesenelfenix.economicas.uba.ar/hoy-mas-que-nunca-es-necesario-formular-e-implementar-politicas-de-empleo-de-calidad

XIV. 2024 LERA Lifetime Achievement Awards

ROBERT CHIARAVALLI
Strategic Labor & Human Resources, LLC

Thank you for this wonderful award. I want to thank the awards committee, as well as Emily Smith, and Bernadette Tiemann. And a special thank-you goes to Dennis Dabney.

When Emily informed me of the award, I was surprised and of course tremendously honored.

I told my wife and children, and they were very congratulatory and happy for me.

The next call I made was to my older brother and older sister. They were also pleased for me, but my sister, with only love in her heart, said, "What have you achieved?" and "Don't people die shortly after a lifetime achievement award?" Both comments properly put my feet back on the ground, where they should be.

But my sister's comments spurred me on to do two things:

- Call my doctor to make sure I was fortifying my health in order to forestall an early demise and expiration; and my estate attorney in the event I could not.
- Consider a retrospective of my professional life.

Regarding my professional life, I have been blessed. I realized the following.

I am doing today what I wanted to do since I was 15 years old. It was the labor relations, the labor law, the conflict and tensions that needed some diplomacy for resolution, and of course the negotiations of labor agreements. It has been a lifelong pursuit to understand it and be a part of it.

My education was geared toward this pursuit. When it came time to decide on either law school or graduate school, I found the ILR program at Cornell. It was a magical place for me. I loved the courses and professors. I had the great blessing to meet and work for Bob McKersie before he left for MIT, and Bob Aronson. The students were great, especially Kathy Fritz, whom I married. More on that in a minute.

Law school came a little later in life, but it was equally wonderful because it made me better at labor relations, and I have been a labor lawyer ever since.

In the course of my business career, I have worked in the research department of the UAW, at the National Labor Relations Board, as a corporate executive and labor counsel, with 20 different international unions, with companies who were heavily unionized, and regulators, all of which allowed me to deal with the challenges of the workplace and working people.

All along the way, LERA (and formerly IRRA) was a constant in my life. I continue to like what it stands for, the relationships, the interesting discussions, and the scholarly research reported here. You are the people I feel most at home with.

Now for my wife, beautiful and bright then and now, accomplished in her own business career and as a partner in our firm for more than 25 years. In love and in life, Kathy and I have been talking about issues of the workplace every day. Amazing!! Her generosity has allowed me to study law and start a private practice. One day while I was at work, I called her and asked if she would mind if I quit and go to law school. Without hesitation, she said, "Go for it." When I was working at an especially challenging startup company, I asked her if she would mind if I quit and start a private practice—with three small kids and a mortgage that could choke a horse—she said, "Go for it."

But the biggest realization, which I never would have considered but for my sister's questions, was that a passionate interest in labor relations and the people and institutions that exist to make sense of them, led to a life of blessings, including this award. An education, a wife, three wonderful children (Elizabeth, Alicia, and Christopher)—all of whom I am proud of daily—without that flame of an interest sparked when I was 15 years old would have never happened.

So, for all of you more junior folks who are building their lifetime of achievements, reach out to us more senior folks. We are dying to help—or put it differently—we would love to help you before we die.

So, in closing, thank you to my LERA colleagues and friends, thank you to the Awards Committee, and thank you to Dennis, Emily and Bernadette.

XIV. 2024 LERA Lifetime Achievement Awards

TERESA GHILARDUCCI
The New School for Social Research

I write these words with great feeling, and I thank LERA president Dennis Dabney and LERA for the Lifetime Achievement Award. I have been a member of the association and Industrial Relations Research Association for 42 years. LERA has been an invaluable community of scholars and practitioners, supporting and inspiring my work throughout my career.

My journey began with a passion for economic justice and a commitment to improving the lives of working people. Early on, I recognized the flaws in the US retirement system, particularly the shift from traditional pension plans to self-managed retirement accounts. This shift left many workers financially insecure in their later years.

I graduated with my B.A. in economics at UC Berkeley at age 20. I had already been working with Lloyd Ulman and Clair Brown at the Institute for Industrial Relations Labor Center, where I was responsible for helping unions in northern California train their negotiators on cost-of living-agreements and pensions. One of those unions was my mother's—one of the few local union newspaper guilds that included the clerks selling classified ads. Years later, I benefited because my mother had a modest amount of income for the rest of her life.

I later founded the Monsignor Higgins Labor Research Center at the University of Notre Dame and the Retirement Equity Lab at The New School. I have supervised 34 Ph.D. students, and they worked for labor, consulting firms, the government, and universities. One such student is now a priest; another is a retired professor and now a novelist. A number of my students have become leaders in the technical departments of trade unions or with progressive economic think tanks.

Over the years, I have had the privilege of collaborating with esteemed colleagues who have enriched my understanding and broadened the impact of my work—Michael Reich, Clair Brown, Chuck Craypo, Frank Wilkinson, Will Milberg, Christian Weller, Anthony Webb, and Jill Rubery.

The forerunners of our labor economists, who are members of LERA, are all a part of the tradition carried on by the neo-institutional labor economists who flourished during and after World War II. These economists advised policymakers and shaped empirical labor economics, following in the footsteps of influential figures such as Clark Kerr, Sumner Slichter, and other presidents of the Industrial Relations Research Association (IRRA), such as John T. Dunlop, Neil W. Chamberlain, and George Shultz. These economists prided themselves on being at the cutting edge of economic theory while being deeply grounded in the institutions that shape power relations and determine economic outcomes.

Institutional labor economics has had a significant presence in major US universities, influencing academic thought and policy development. Many presidents of IRRA have also served as presidents of major universities such as Harvard and the University of Chicago, underscoring the close ties between academic leadership and the development of labor economics in the United States.

I am grateful to the bricklayers, steelworkers, and operating engineers unions, as well as UAW, CWA, and AFL-the CIO. I also am grateful to Randi Weingarten, Richard Trumka, Tom Donahue, Liz Schuler, Jack Joyce, four UAW presidents, George Becker, Leo Gerard, and Andy Stern.

One of my notable achievements is the development of guaranteed retirement accounts (GRAs), a proposal designed to ensure all workers have access to secure and sufficient retirement savings. This innovative idea has garnered significant attention and support, exemplifying my commitment to creating a fair and just economic system.

Receiving this award from LERA is a testament to the power of community and collaboration. I am deeply grateful for the support and inspiration I have received from this remarkable organization. Together, we have made significant strides in ensuring that all workers have the opportunity to achieve financial security and retire with dignity.

XV. LERA Annual Reports

LERA Executive Board Meeting Minutes
Monday, February 12, 2024, 10:30 a.m. – 12:45 p.m. ET
Via Videoconference

Call to Order—The meeting was called to order at 10:30 a.m. by Paul Clark, President. Present at the meeting were board members Peter Berg, Andrea Cáceres, Virginia Doellgast, Robert Chiaravalli (RVP Mid), Janet Gillman, Beverly Harrison (RVP East), John Johnson, Tamara Lee, Keven Legel, Deb Mueller, Dionne Pohler, Javier Ramirez, Chrisine Riordan, Sean Rogers, and Christy Yoshitomi. Committee Chairs present were Bonnie Castrey (NCAC Vice Chair, Media Award Chair), Joel Cutcher-Gershenfeld (IC/IS Chair), Virginia Doellgast (Diversity and Inclusion Chair), and David Lewin (Membership Committee Co-Chair). Officers present were Paul Clark (President), Bill Canak (NCAC), Ryan Lamare (Editor-in-Chief), and Andrew Weaver (Secretary–Treasurer). Staff present were Emily Smith and Bernadette Tiemann.

Approval of the Minutes—The motion to approve the minutes of the June 2023 board meeting passed unanimously with no revisions. David Lewin moved and William Canak seconded the motion to approve the minutes, and it unanimously passed. The General Membership Meeting minutes from 2023 were brought to the floor, and any corrections were asked for. The board passed a motion unanimously to approve the minutes, the motion was made by Quinton Herbert and seconded by David Lewin.

Reports

Financial Report—Andrew Weaver, LERA Secretary–Treasurer, reports a strong financial outcome in 2023 with top-line growth of $588,000 up from $450,000 in 2022. LERA continues to have major sustaining sponsorships from KP, Ford, and UAW. In terms of net income and bottom-line performance, LERA had a 14% net income for the year. In the past five years, LERA has held to a steady 11–14% net income range, except for 2020, which was higher. The cost structure for LERA is similar year over year, and this continuity and stability is good to see.

In terms of items that are increasing for the coming year, the big-ticket item is the New York conference in June 2024, which combines three major events, including the LERA 76th Annual Meeting, the ILERA 20th Triennial World Congress, and the Federal Mediation and Conciliation Service's National Labor–Management Conference. Because of the scale of this event, and the fact that LERA is managing the financials, the organization will see a large increase both in expenses and revenues. We do expect to clear a decent margin from this combined event, given that LERA will be keeping registration fees, and this lays the foundation for LERA to consider a third staff member to support its programs.

The second area of increased expense in 2024 is a planned increase of five percent increase in staff salaries. Most of this increase reflects inflation in cost of living, and there is also a component to recognize staff merit. Both the LERA Executive Committee and the LERA Executive Board support this increase.

From a balance sheet perspective, LERA has a defensive interval of about 18 months of operations, which is a decent position to be in. In 2024, we are increasing the scale of operations, but we are also focusing on preserving financial stability. The LERA Executive Board approved the year-end financials for 2023 as well as the 2024 operating budget.

Jim Pruitt remarked that he is about to go into his year as president, and he has big shoes to fill. His commitment to the Executive Board and the organization is to focus on increasing membership. He is happy to be working with team members Emily Smith and Bernadette Tiemann on staff as well as Andrew Weaver as Secretary–Treasurer of the organization.

Membership Committee Report—David Lewin reports that we raised dues a year ago, and we do not plan to do so this year. In terms of the makeup of our membership, we are over 50% academics, management is about 10%, and union is about 11%.

LERA Annual Meeting Program Committee Report—Harry Katz reports that the meeting is planned, and there was substantial interest in participating in the program, so the meeting will actually happen between Wednesday, June 26, 2024, through Sunday, June 30, 2024. We are impressed with the number of internationals who plan to attend, and the organization is financially supporting a number of these attendees who could not otherwise afford to attend. Staff is working miraculously on developing the meeting. The registration fees have been kept to $385 for early bird member registration, the same as in Detroit. This fee is substantially less than what FMCS has charged in the past, and in the range of what ILERA has charged for their past meetings. Jim Pruitt thanks Dennis Dabney, Bill Canak, and FMCS for their assistance on the program.

David Lewin reports a record high number of submissions for the LERA/AILR Best Papers, most of them international submissions.

We have also raised money to help support the Triad. Universities have provided sponsorship funds and some sponsorships of individual sessions; this money is around $106,000 at this point, and we also know that the FMCS is just now beginning their own efforts to raise sponsorship funds, so this will really help attain a positive revenue stream from the conference.

Editorial Committee Report—Ryan Lamare reports that the 2024 Research Volume, *Union Organizing and Collective Bargaining at a Critical Moment*, edited by Howard Stanger, Paul Clark, and John Delaney, is moving through the stages of production. We have a few outstanding chapters that are wrapping up in the next week or two, but we are nicely on track for an on-time release of this volume. The 2025 research volume is also in the works. The proposal will be on the next board meeting's agenda, and it will be circulated to the board for approval in the next few weeks. It will focus on the role of labor and employment in the climate crisis. Richard Benton, University of Illinois, is currently working on the proposal with a second editor, Lara Skinner, Cornell University. They have several chapter authors lined up, and we should be getting a formal proposal in the next few weeks.

Interest Sections/Industry Councils Coordinating Committee Report—Bernadette Tiemann reports that the Industry Councils and Interest Sections have been reworking and rebranding the areas of the various groups and revitalizing the leadership. The Teaching Interest Section has attracted large amounts of attention. The group hopes to offer virtual content to registered attendees in the leadup to the event. Lastly, there have been three new groups chartered—hospitality, sports and entertainment, and social impact organizations. Janet Gillman reported on reviving the Public Sector Industry Council and is currently locating leadership for this council, surveying members to get an understanding of what they are interested in and potentially putting on a program for future meetings. Jim Pruitt asked about putting The Hospitality Industry Council leaders into contact with him because knows of interested parties in Hawaii UNITE-HERE. Andrew Weaver reports that the Industry Studies Association (ISA) has had some ups and downs in the last few years, and it might be beneficial for both organizations to have closer ties, perhaps even one day have our staff support their organization. Anyone with connections to the ISA is encouraged to support closer connections and ties. David Lewin said that the Industry Councils and Interest Sections report has some conflicting information. Railroads and Trucking has been combined, as well as some others. Peter Berg mentioned that ISA ran a combined meeting in 2015 in Portland, and the overlapping meetings worked pretty well. He supports the idea of meeting together in the future.

Joel Cutcher-Gershenfeld reports that Bernadette Tiemann has been an amazing partner on the councils and sections. Joel is recommending each of the groups to target putting on one webinar in the leadup to the meetings in Manhattan. Joel proposed that we rebrand Work and Employment Relations to become Work, Human Resources, and Organizations, that the Collective Bargaining Network become Collective Bargaining, and the Dispute Resolution (Labor and Employment Law) become Dispute Resolution, Labor and Employment Law. The motion for these name changes was brought by Joel Cutcher-Gershenfeld, and the motion was seconded by Jim Pruitt. The motion passed unanimously.

National Chapter Advisory Council Report—Bill Canak reports that the NCAC met last month in a well-attended meeting. Each of the RVPs provided an overview of their region. Most chapters have revived from the pandemic challenges, and their finances look to be in good shape. We have new chapters forming in Minnesota, Dallas, Miami, Buffalo, and Las Vegas. We had a successful revival of the Capital Region New York Chapter. We have had some discussion of chapter bylaws and encouraged some revision that would make administration of the bylaws more efficient. We solicited and are expecting to hear nominations for the next Mid Region RVP, and we thank Bob Chiaravalli for the wonderful job he has done in this role. The NCAC voted to support a chapter table at the upcoming Manhattan meeting; there should be many people there who have not been to a LERA conference before, and this will be way to provide information and answer questions about local LERA chapters. Bill Canak reminded the board that it has asked that all committee chairs and committee members join their local chapter. He requested that everyone fulfill this requirement, and he reminded the group that all LERA national members have been required to become chapter members since 2016. The NCAC encourages chapters to submit session proposals to the Program Committee. Dennis Dabney also reminded people that they can join the virtual chapter of LERA to fulfill this requirement.

Bill Canak would like to consider the Central Region and asked that anyone who has thoughts about who would be appropriate for the Mid RVP to please share their name and contact information with Bill Canak or Bonnie Castrey.

Development Committee Report—Harry Katz reports that the development process has gone well in 2023. We had a continuing solid annual fund contributions in the amount of $75,000, and $75,000 in major sustaining sponsors (Ford, KP, and UAW). Organizational memberships have also gone up, and this had continued to drift upward in the last several years. We continue to work at building up an endowment; there are a few bequest gifts that have been made to LERA. Katz asked everyone to please keep this in mind in their own estate planning.

Jim Pruitt reports that since Dennis Dabney has retired, we have lost an important part of support from KP. Fortunately, Hal Ruddick is also on the board, and Hal is picking up this support from LERA. Between Hal Ruddick and Jim Pruitt, KP continues its support of LERA.

New Business

No new business.

Adjournment—Hearing no new business, the meeting was adjourned at 12:29 pm ET by President Dennis Dabney. Motion to adjourn by William Canak, seconded by Jim Pruitt, and passed unanimously. *The Executive Board will meet next in June 2024 in Manhattan in conjunction with the annual meeting.*

LERA Executive Board Meeting Minutes
Friday, June 28, 2024, 4:30–6 p.m. ET
Hilton Midtown Manhattan, Rendezvous Trianon, 1335 Avenue of the Americas, New York, NY 10019

Call to Order—The meeting was called to order at 4:30 p.m. by Dennis Dabney, President. Present at the meeting were officers Dennis Dabney (President), Paul Clark (Past President), Jim Pruitt (President-Elect), Ryan Lamare (Editor-in-Chief), and Andrew Weaver (Secretary–Treasurer). Board members present were Peter Berg, Andrea Cáceres, Janet Gillman, Dionne Pohler, Sean Rogers, Marc Weinstein (DCDC Chair), and Christy Yoshitomi. Regional Vice Presidents present were Robert Chiaravalli (Mid), Beverly Harrison (East), and Jim Pruitt (West). Committee Chairs present were John Budd (Nominating Committee Chair) Bonnie Castrey (NCAC Vice Chair), Joel Cutcher-Gershenfeld (Community Councils, Media Award Chair), Elena Falcettoni (LERA@ASSA Program Committee Chair), Taz Hurst (Membership Committee Co-Chair), David Lewin (Membership Committee. Co-Chair), and Lionel Sims Jr. (Diversity and Inclusion Committee Chair). Staff present were Emily Smith and Bernadette Tiemann.

Approval of the Minutes—The minutes of the LERA Executive Board meeting in February 2024 were reviewed by the board and unanimously approved.

Reports

Financial Report—Andrew Weaver, LERA Secretary–Treasurer, reports that LERA achieved a net balance of about 14% in both 2023 and 2022. The key takeaway for 2024 is that this meeting positions LERA to alter its trajectory. This high-risk, high-reward moment will result in a healthy margin, allowing for the addition of a third staff position. This new role is essential for the organization's growth, and the funding from this meeting enables it. Over the next four years, LERA has a window to boost membership to sustain a staff member at approximately $80,000 per year. This meeting marks a transformational step, with a clear plan in place for leveraging the 2024 funds to elevate the organization. The new staff position will be contractual.

Membership Committee Report— David Lewin, Membership Committee Co-Chair, reports that LERA has added 240 new members in the past six months, with significant growth in academic members and neutrals. While the number of consultants and attorneys has halved since 2010, the committee aims to attract more management members. Some initiatives are under way, including potential online programs offering credits to practitioner members as a membership benefit. Reflecting on LERA's independent meeting since June 2013, David emphasized that the organization has sustained its financial model over the last decade.

Development Committee Report—Jim Pruitt, Development Committee Co-Chair, reports that the Development Committee includes himself, Harry Katz, Peter Berg, Jennifer Werner, and Shankar Vis. In 2023, Paul Clark raised funds to support staff salaries. For 2024, LERA expects sponsorships from Kaiser Permanente, UAW, Ford, and possibly El Camino Health. The FMCS, by law, does not accept profits, but it partnered with LERA for the Triad 2024 conference, which proved profitable. Jim encouraged Executive Board members to consider making a donation to LERA, proposing a contribution of $1,000, although any amount would be appreciated.

77th Annual Meeting Program Committee Report— John Budd reports that the theme of the 2025 Annual Meeting is "Promoting Authentic Dialogue in Polarized Times." Erin Kelly from MIT will serve as Academic Vice Chair, while Beth Schindler of FMCS will serve as Practitioner Chair. Connections with Microsoft and Boeing are confirmed, and the event will be held in Seattle June 12–15, 2025.

Nominating Committee—John Budd, Committee Chair, reports on the slate for the 2024 election. The committee recommends Beverly Harrison for President in 2025-26, along with other candidates listed in the slate. Paul Clark moved to accept the slate, seconded by Christy Yoshitomi. The slate passed unanimously.

Diversity and Inclusion Committee—Lionel Sims, Jr., Committee Chair, reports on new committee members: Asha Alt, Frank Mullins, Kevin Stokes, and Sarah Cudahy. The committee discussed membership demographics and meeting registration trends to establish baseline data for future analysis. They proposed surveying members and attendees to gauge their experiences at meetings and whether they felt welcomed. Additionally, the committee is considering creating a DEI interest group to allow members to engage more directly with these issues.

Editorial Committee Report— Ryan Lamare, LERA Editor-in-Chief, reports that the 2024 Research Volume, *Union Organizing and Collective Bargaining at a Critical Moment* (edited by Howard Stanger, Paul Clark, and John Delaney) is in its final stages of production and will be circulated soon. The 2025 Research Volume, *The Climate-Labor Movement: Lessons Learned and the Promise of Equitable and Diverse Clean Energy Economy*, edited by Richard Benton and Lara Skinner, is on track for production in 2025. The committee discussed ways to streamline the process for future volumes, with plans to present a proposal for the 2026 volume in the next six months. The *LERA/ILR Review* continues to accept papers from the annual meetings. Ryan announced that he would transition out of his role as Editor-in-Chief, with Helen Liu set to take over. Helen, a longtime LERA member and professor at Rutgers University, will officially assume the position soon. Christy Yoshitomi moved to approve the 2025 Research Volume, seconded by Janet Gillman. The motion passed unanimously.

National Chapter Advisory Council Report—Bonnie Castrey, NCAC Vice Chair, reports that the NCAC has held three meetings since the conference began. Chapters are gradually recovering from the pandemic, with discussions on bylaws and sustainable leadership models for local officers. Regional Vice Presidents have been instrumental in supporting local chapters, and the national staff has facilitated online meetings. New chapters have emerged in Florida, North Star, and Capital Region NY. The idea of consolidating national LERA dues and chapter dues on a single invoice was discussed, and staff will explore this with the accountant and Secretary–Treasurer.

Community Councils Coordinating Committee Report—Joel Cutcher-Gershenfeld, Chair, reports that new charters for three news councils—Social Impact, Hospitality, and Climate Jobs—have been approved. These councils, once funded by the Sloan Foundation, continue to thrive after the funding has been exhausted for many years. The committee is exploring a potential DEI interest group, which would become the fifth community under their guidance. The board also discussed changing the terminology from "industry councils" and "interest sections" to "community councils," each with its own governing council. Three new co-chairs will succeed Joel in his coordinating role: Christy Yoshitomi, Kevin Hawkins, and Mark Gough.

LERA@ASSA Program Committee—Elena Falcettoni, Committee Chair, reports that the 2024 sessions were lively, with Brad Hershbein co-chairing the program. Attendance at ASSA has decreased to about half of pre-COVID levels. The committee is planning 18 sessions for the January 2025 conference in San Francisco, despite the general attendance decline. Elena Falcettoni will be joined by Illenin Kondo (Federal Reserve Bank) in co-chairing the San Francisco meeting in January 2025.

New Business

LERA Online LMS—Antone Aboud proposed creating online learning content to support practitioner membership benefits. Classes would be offered both synchronously and asynchronously, with some small investment costs, such as renting mailing lists. Dennis Dabney requested input from the Membership Committee and DCDC to assess interest. If positive, a feasibility group should be formed to explore compatibility with university offerings. The board noted that while LERA should avoid competing with universities, it could play a central role in administering competency frameworks alongside university programs. Antone will collaborate with Joel Cutcher-Gershenfeld and

Bob Chiaravalli to incorporate past work from 2009, create a financial model with DCDC, and present a proposal to the board.

Adjournment—The meeting adjourned at 6:15 p.m. ET.

LERA General Membership Meeting Minutes
Saturday, June 29, 2024, 4:45–6 p.m. ET
Hilton Midtown Manhattan, Trianon Jr. Ballroom, 1335 Avenue of the Americas, New York, NY 10019

Call to Order—President Dennis Dabney called the meeting to order and welcomed new LERA members. He introduced Jim Pruitt, President-Elect, who will assume the role of LERA President at the end of the meeting.

Reports

Report on 2023 Election Outcome and Upcoming Election—Dennis Dabney announced the results of the 2023 election and highlighted the upcoming election scheduled for later this summer. New board members Maite Tapia, Maria Figueroa, Monique Morrissey, Hal Ruddick, Lionel Sims, and Janet Gillman were recognized. John Budd was introduced as the 2024-25 LERA President- Elect.

Finance Report—Secretary–Treasurer Andrew Weaver reported strong financials for 2023, noting that the organization has an 18-month financial cushion entering 2024.

Membership Committee Report—David Lewin reported that membership grew from 1,039 in 2023 to 1,266 in 2024 due to active promotion efforts. Future membership initiatives are in progress.

Development Committee Report—Jim Pruitt thanked Andrew Weaver for his dedicated service as Secretary–Treasurer and for Past President Paul Clark's significant fundraising efforts that led to increased staff salaries. The UAW, Ford, and KP renewed their sponsorships. Members were encouraged to contribute to LERA.

Editorial Committee Report—Ryan Lamare announced the 2024 annual research volume, "Union Organizing and Collective Bargaining," edited by Paul Clark and Howard Stanger. The 2025 edition, "The Climate Labor Movement," will be edited by Richard Benton and Lara Skinner.

Program Committee Reports—

- LERA 76th Annual Meeting (2024): Jim Pruitt, Harry Katz, and Javier Ramirez reported that Triad 2024 was successful financially and provided valuable networking opportunities. Ramirez emphasized blending theory and practice through collaboration.
- LERA 77th Annual Meeting (2025): John Budd announced that the next meeting will take place June 12–15, 2025, in Seattle, WA. He will co-chair with Erin Kelly and Beth Schindler. The theme is "Promoting Authentic Dialogue in Polarized Times." The session proposal deadline is November 15, 2024.
- LERA@ASSA 2025: Elena Falcettoni provided an update on the upcoming event in San Francisco, January 3–5, 2025.

Diversity and Inclusion Committee Report—Lionel Sims, Jr. reported that four new members joined the committee. The committee is exploring the creation of a DEI interest group for member engagement.

National Chapter Advisory Council Report—Bonnie Castrey reported that chapters are recovering post-COVID and are resuming in-person meetings. New chapters were established in Dallas, Minnesota, and New York. She praised the LERA staff for their support in membership growth and digital resources.

Industry Council and Interest Sections Report—Joel Cutcher-Gershenfeld announced the formation of a new Social Impact Council, Hospitality/Sports/Entertainment group, and a Climate Jobs/Green Jobs group. The DEI Committee is also working to launch a DEI Council. All groups will now be referred to as "Community Councils."

Awards Ceremony

- Kenneth May Media Awards: Presented by David Lewin and Bonnie Castrey to Heather Cox Richardson and Derek Seidman
- Thomas Kochan and Stephen Sleigh Best Dissertation Award: Presented by Dionne Pohler to John Kallas (Cornell University) for "Bringing the Strike Back In: Understanding Strike Strategies and Effectiveness in the United States."
- DCDC Best Student Paper Award: Presented by Marc Weinstein to Cherise Regier for "Digital Overconnectivity and Employee Wellbeing: An Evaluation of the Right-to-Disconnect Policy in Europe."
- James G. Scoville Best International/Comparative IR/HR Paper Awards: Presented by Marc Weinstein to Jonathan Preminger and Assef Bondy, and Nick Krachler.
- John T. Dunlop Outstanding Scholar Awards: Presented by Rose Batt to Tingting Zhang and Hatim Rahman.
- Outstanding Practitioner Award: Presented to Javier Ramirez.
- LERA Fellows Awards: Academic Fellows—Peter Berg, Stephen Silvia, Paul Clark, and Laura Dresser. Practitioner Fellows—Michael LeRoy, Hal Ruddick, Beverly Harrison, and Jeanne Charles.
- Susan C. Eaton Outstanding Scholar–Practitioner Award: Presented to Terri Gerstein.
- Myron T. Taylor Management Award: Presented to Jim Pruitt.
- LERA Chapter Star Award: Presented by Bonnie Castrey to NYC LERA.
- Lifetime Achievement Award: Presented by Dennis Dabney to Bob Chiaravalli.

New Business

Bonnie Castrey thanked Dennis Dabney for serving a second term as LERA President.

Announcements

Dennis Dabney announced the dates of upcoming LERA meetings: LERA 77th Annual Meeting: June 12–15, 2025, in Seattle, WA; and LERA@ASSA 2025: January 3–5, 2025, in San Francisco, CA

Closing Remarks and Adjournment

Dennis Dabney ceremonially handed the gavel to Jim Pruitt, who officially assumed the LERA presidency. Jim Pruitt presented Dennis Dabney with a plaque in recognition of his service.

The meeting was adjourned at 6:00 PM by President Jim Pruitt.

Audit Reports for 2022 and 2023

Labor and Employment Relations Association

Champaign, Illinois

Financial Statements

For the Years Ended December 31, 2023 and 2022

Feller & Kuester CPAs PLLC
Certified Public Accountants
806 Parkland Court
Champaign, IL 61821
217-351-3192

TABLE OF CONTENTS

Feller & Kuester CPAs PLLC

Tax - Audit - Bookkeeping
806 Parkland Court - Champaign, Illinois 61821
Phone - (217) 351-3192 Fax - (217) 351-4135 Email - admin@fellerkuester.com

INDEPENDENT AUDITOR'S REPORT

To the Board of Directors of
Labor and Employment Relations Association
Champaign, Illinois

Opinion

We have audited the accompanying financial statements of the Labor and Employment Relations Association (a nonprofit organization), which comprise the statements of financial position as of December 31, 2023 and 2022, and the related statements of activities, functional expenses, and cash flows for the years then ended, and the related notes to the financial statements.

In our opinion, the financial statements referred to above present fairly, in all material respects, the financial position of the Labor and Employment Relations Association as of December 31, 2023 and 2022, and the changes in its net assets and its cash flows for the years then ended in accordance with the accounting principles generally accepted in the United States of America.

Basis for Opinion

We conducted our audits in accordance with auditing standards generally accepted in the United States of America. Our responsibilities under those standards are further described in the Auditor's Responsibilities for the Audit of the Financial Statements section of our report. We are required to be independent of the Labor and Employment Relations Association and to meet our other ethical responsibilities in accordance with the relevant ethical requirements relating to our audits. We believe that the audit evidence we have obtained is sufficient and appropriate to provide a basis for our audit opinion.

Responsibilities of Management for the Financial Statements

Management is responsible for the preparation and fair presentation of the financial statements in accordance with accounting principles generally accepted in the United States of America, and for the design, implementation, and maintenance of internal control relevant to the preparation and fair presentation of financial statements that are free from material misstatement, whether due to fraud or error.

In preparing the financial statements, management is required to evaluate whether there are conditions or events, considered in the aggregate, that raise substantial doubt about the Labor and Employment Relations Association's ability to continue as a going concern within one year after the date that the financial statements are available to be issued.

Auditor's Responsibilities for the Audit of the Financial Statements

Our objectives are to obtain reasonable assurance about whether the financial statements as a whole are free from material misstatement, whether due to fraud or error, and to issue an auditor's report that includes our opinion. Reasonable assurance is a high level of assurance but is not absolute assurance and therefore is not a guarantee that an audit conducted in accordance with generally accepted auditing standards will always detect a material misstatement when it exists. The risk of not detecting a material misstatement resulting from fraud is higher than for one resulting from error, as fraud may involve collusion, forgery, intentional omissions, misrepresentations, or the override of internal control. Misstatements are considered material if there is a substantial likelihood that, individually or in the aggregate, they would influence the judgment made by a reasonable user based on the financial statements.

In performing an audit in accordance with generally accepted auditing standards, we:

- Exercise professional judgment and maintain professional skepticism throughout the audit.
- Identify and assess the risks of material misstatement of the financial statements, whether due to fraud or error, and design and perform audit procedures responsive to those risks. Such procedures include examining, on a test basis, evidence regarding the amounts and disclosures in the financial statements.
- Obtain an understanding of internal control relevant to the audit in order to design audit procedures that are appropriate in the circumstances, but not for the purpose of expressing an opinion on the effectiveness of the Labor and Employment Relations Association's internal control. Accordingly, no such opinion is expressed.
- Evaluate the appropriateness of accounting policies used and the reasonableness of significant accounting estimates made by management, as well as evaluate the overall presentation of the financial statements.
- Conclude whether, in our judgment, there are conditions or events, considered in the aggregate, that raise substantial doubt about the Labor and Employment Relations Association's ability to continue as a going concern for a reasonable period of time.

We are required to communicate with those charged with governance regarding, among other matters, the planned scope and timing of the audit, significant audit findings, and certain internal control related matters that we identified during the audit.

Feller & Kuester CPAs PLLC

Feller & Kuester CPAs PLLC
Champaign, Illinois

November 13, 2024

LABOR AND EMPLOYMENT RELATIONS ASSOCIATION
STATEMENTS OF FINANCIAL POSITION
DECEMBER 31, 2023 AND 2022

	2023	2022
ASSETS		
Current Assets		
Cash and Cash Equivalents	$ 643,422	$ 435,876
Investments	71,125	81,978
Accounts Receivable, Net	12,609	1,773
Prepaid Expenses	30,661	18,219
Inventory	5,418	7,228
Total Current Assets	763,235	545,074
Property and Equipment, Net	1,101	1,630
TOTAL ASSETS	$ 764,336	$ 546,704
LIABILITIES AND NET ASSETS		
Current Liabilities		
Accounts Payable	$ 762	$ 250
Accrued Liabilities	48,778	45,077
Dues Collected in Advance	82,119	84,283
Subscriptions Collected in Advance	9,082	12,716
Deferred Chapter Dues	4,300	2,850
Other Deferred Revenue	154,005	10,450
Total Current Liabilities	299,046	155,626
Net Assets		
Without Donor Restrictions		
Designated by the Board		
Susan C. Eaton Scholar-Practitioner Memorial Fund	12,779	16,779
Kochan-Sleigh Best Dissertation Award Fund	34,200	35,319
Jim Scoville Award Fund	818	2,176
John T. Dunlop Public Policy Fund	18,127	16,503
Gladys and Water Gershenfeld Diversity & Inclusion Fund	19,494	17,939
Quasi-Endowment Fund	35,462	45,373
Total Designated by the Board	120,880	134,089
Undesignated	307,226	222,085
Total Without Donor Restrictions	428,106	356,174
With Donor Restrictions	37,184	34,904
Total Net Assets	465,290	391,078
TOTAL LIABILITIES AND NET ASSETS	$ 764,336	$ 546,704

See Accompanying Notes.

LABOR AND EMPLOYMENT RELATIONS ASSOCIATION
STATEMENT OF ACTIVITIES
FOR THE YEAR ENDED DECEMBER 31, 2023

	Without Donor Restrictions			With Donor	
	Undesignated	Designated	Total	Restrictions	Total
Support and Revenues					
Membership Dues	$ 144,716	$ -	$ 144,716	$ -	$ 144,716
Meeting Income	224,532	-	224,532	1,830	226,362
Organization Dues & Sponsorships	95,608	-	95,608	450	96,058
Subscriptions	14,872	-	14,872	-	14,872
Chapter Fees	9,180	-	9,180	-	9,180
Member Contributions	75,840	1,827	77,667	-	77,667
Royalties	5,024	-	5,024	-	5,024
Ad Income	11,630	-	11,630	-	11,630
Publications	1,614	-	1,614	-	1,614
Administrative Fees	1,225	-	1,225	-	1,225
Investment Income	(2,477)	(8,376)	(10,853)	-	(10,853)
Interest Income	421	-	421	-	421
Net Assets Released from Designation	6,660	(6,660)	-	-	-
Total Support and Revenues	588,845	(13,209)	575,636	2,280	577,916
Expenses					
Program Services					
General	220,649	-	220,649	-	220,649
Meetings	138,568	-	138,568	-	138,568
Publications	40,491	-	40,491	-	40,491
Supporting Services					
Management and General	76,306	-	76,306	-	76,306
Membership Development	27,690	-	27,690	-	27,690
Total Expenses	503,704	-	503,704	-	503,704
Change in Net Assets	85,141	(13,209)	71,932	2,280	74,212
Net Assets, Beginning of Year	222,085	134,089	356,174	34,904	391,078
Net Assets, End of Year	$ 307,226	$ 120,880	$ 428,106	$ 37,184	$ 465,290

See Accompanying Notes.

LABOR AND EMPLOYMENT RELATIONS ASSOCIATION
STATEMENT OF ACTIVITIES
FOR THE YEAR ENDED DECEMBER 31, 2022

	Without Donor Restrictions			With Donor	
	Undesignated	**Designated**	**Total**	**Restrictions**	**Total**
Support and Revenues					
Membership Dues	$ 150,132	$ -	$ 150,132	$ -	$ 150,132
Meeting Income	115,145	-	115,145	5,888	121,033
Organization Dues & Sponsorships	94,227	-	94,227	550	94,777
Subscriptions	14,121	-	14,121	-	14,121
Chapter Fees	10,510	-	10,510	-	10,510
Member Contributions	48,299	3,775	52,074	-	52,074
Royalties	6,257	-	6,257	-	6,257
Ad Income	9,130	-	9,130	-	9,130
Administrative Fees	1,195	-	1,195	-	1,195
Investment Income	(22)	(3,955)	(3,977)	-	(3,977)
Interest Income	139	-	139	-	139
Net Assets Released from Designation	6,528	(6,118)	410	(410)	-
Total Support and Revenues	455,661	(6,298)	449,363	6,028	455,391
Expenses					
Program Services					
General	236,920	-	236,920	-	236,920
Meetings	2,709	-	2,709	-	2,709
Publications	40,516	-	40,516	-	40,516
Supporting Services					
Management and General	70,967	-	70,967	-	70,967
Membership Development	27,658	-	27,658	-	27,658
Total Expenses	378,770	-	378,770	-	378,770
Change in Net Assets	76,891	(6,298)	70,593	6,028	76,621
Net Assets, Beginning of Year	145,194	140,387	285,581	28,876	314,457
Net Assets, End of Year	$ 222,085	$ 134,089	$ 356,174	$ 34,904	$ 391,078

See Accompanying Notes.

LABOR AND EMPLOYMENT RELATIONS ASSOCIATION
STATEMENT OF FUNCTIONAL EXPENSES
FOR THE YEAR ENDED DECEMBER 31, 2023

		MEETINGS			PUBLICATIONS			SUPPORTING SERVICES		
	General	Annual Conference	ASSA Meeting	Other Meetings	Perspectives	Research Volume	Other Publications	Management & General	Membership Development	Total
Compensation	$ 154,611	$ -	$ -	$ -	$ -	$ -	$ -	$ 29,076	$ -	$ 183,687
Payroll Taxes & Fringes	42,576	-	-	-	-	-	-	8,004	-	50,580
Contract Labor	23,462	-	-	-	-	-	-	-	-	23,462
Depreciation	-	-	-	-	-	-	-	529	-	529
Insurance	-	-	-	-	-	-	-	1,496	-	1,496
Bank and Service Charges/Fulfillment	-	-	-	-	-	-	-	20,577	-	20,577
Promotion	-	1,805	-	-	-	-	-	-	7,266	9,071
Postage/Freight	-	-	-	8	1,543	1,710	-	238	-	3,499
Accounting/Auditing	-	-	-	-	-	-	-	8,746	-	8,746
Printing and Production	-	2,339	116	-	11,733	6,842	-	-	-	21,030
Services	-	814	-	-	6,500	5,882	6,269	-	-	19,465
Other Publication Costs	-	-	-	-	12	-	-	-	-	12
Meals/Receptions	-	73,294	5,708	-	-	-	-	-	-	79,002
Travel	-	3,855	2,318	-	-	-	-	-	-	6,173
Other Meeting Expenses	-	47,873	438	-	-	-	-	-	-	48,311
Computer Supplies/Services	-	-	-	-	-	-	-	836	-	836
Website	-	-	-	-	-	-	-	2,793	-	2,793
Office Supplies	-	-	-	-	-	-	-	4,003	-	4,003
Student & Member Awards	-	-	-	-	-	-	-	-	9,256	9,256
Fundraising Expense	-	-	-	-	-	-	-	-	686	686
Chapter Expenses	-	-	-	-	-	-	-	-	10,286	10,286
Other Committee Expenses	-	-	-	-	-	-	-	-	196	196
Miscellaneous Office	-	-	-	-	-	-	-	8	-	8
Total	$ 220,649	$ 129,980	$ 8,580	$ 8	$ 19,788	$ 14,434	$ 6,269	$ 76,306	$ 27,690	$ 503,704

See Accompanying Notes.

LABOR AND EMPLOYMENT RELATIONS ASSOCIATION
STATEMENT OF FUNCTIONAL EXPENSES
FOR THE YEAR ENDED DECEMBER 31, 2022

		MEETINGS			PUBLICATIONS			SUPPORTING SERVICES		
	General	Annual Conference	ASSA Meeting	Other Meetings	Perspectives	Research Volume	Other Publications	Management & General	Membership Development	Total
Compensation	$ 147,019	$ -	$ -	$ -	$ -	$ -	$ -	$ 30,358	$ -	$ 177,377
Payroll Taxes & Fringes	59,382	-	-	-	-	-	-	11,445	-	70,827
Contract Labor	30,519	-	-	-	-	-	-	-	-	30,519
Depreciation	-	-	-	-	-	-	-	1,662	-	1,662
Insurance	-	-	-	-	-	-	-	1,261	-	1,261
Bank and Service Charges/Fulfillment	-	-	-	-	-	-	-	12,733	-	12,733
Promotion	-	-	-	-	-	-	-	-	12,206	12,206
Postage/Freight	-	-	-	-	2,031	2,602	-	407	-	5,040
Accounting/Auditing	-	-	-	-	-	-	-	7,038	-	7,038
Printing and Production	-	-	614	-	12,229	7,757	-	-	-	20,600
Services	-	149	-	-	5,580	10,046	-	-	-	15,775
Other Publication Costs	-	-	-	-	78	-	193	-	-	271
Meals/Receptions	-	73	-	-	-	-	-	-	-	73
Travel	-	1,128	-	425	-	-	-	-	-	1,553
Other Meeting Expenses	-	65	255	-	-	-	-	-	-	320
Computer Supplies/Services	-	-	-	-	-	-	-	1,101	-	1,101
Website	-	-	-	-	-	-	-	133	-	133
Office Supplies	-	-	-	-	-	-	-	4,577	-	4,577
Student & Member Awards	-	-	-	-	-	-	-	-	7,587	7,587
Fundraising Expense	-	-	-	-	-	-	-	-	1,683	1,683
Chapter Expenses	-	-	-	-	-	-	-	-	6,182	6,182
Miscellaneous Office	-	-	-	-	-	-	-	252	-	252
Total	$ 236,920	$ 1,415	$ 869	$ 425	$ 19,918	$ 20,405	$ 193	$ 70,967	$ 27,658	$ 378,770

See Accompanying Notes.

LABOR AND EMPLOYMENT RELATIONS ASSOCIATION
STATEMENTS OF CASH FLOWS
FOR THE YEARS ENDED DECEMBER 31, 2023 AND 2022

	2023	2022
CASH FLOWS FROM OPERATING ACTIVITIES		
Change in Net Assets	$ 74,212	$ 76,621
Adjustments to Reconcile Change in Net Assets to Net Cash Provided by (Used in) Operating Activities:		
Depreciation Expense	529	1,662
Net Unrealized (Gains) Losses on Investments	14,636	6,756
(Increase) Decrease in Operating Assets:		
Accounts Receivable, Net	(10,836)	(367)
Prepaid Expenses	(12,442)	(12,593)
Inventory	1,810	(856)
Increase (Decrease) in Operating Liabilities:		
Accounts Payable	512	250
Accrued Liabilities	3,701	5,416
Dues Collected in Advance	(2,164)	8,366
Subscriptions Collected in Advance	(3,634)	4,879
Deferred Chapter Dues	1,450	(2,350)
Other Deferred Revenue	143,555	(44,100)
Net Cash Provided by (Used in) Operating Activities	211,329	43,684
CASH FLOWS FROM INVESTMENT ACTIVITIES		
Reinvested Interest, Dividends, and Capital Gains	(3,783)	(2,779)
Purchases of Property and Equipment	-	-
Net Cash Provided by (Used in) Investment Activities	(3,783)	(2,779)
CASH FLOWS FROM FINANCING ACTIVITIES		
None	-	-
Net Cash Provided by (Used in) Financing Activities	-	-
Net Increase (Decrease) in Cash and Cash Equivalents	207,546	40,905
Beginning Cash and Cash Equivalents	435,876	394,971
Ending Cash and Cash Equivalents	$ 643,422	$ 435,876

See Accompanying Notes.

NOTE 1 – NATURE OF ACTIVITIES AND SIGNIFICANT ACCOUNTING POLICIES

Organization and Nature of Activities

The Labor and Employment Relations Association (the Organization) was founded in 1947 to encourage research in all aspects of the field of labor, employment, and the workplace. It is a nonprofit scholarly association of academic, labor, business, and neutral communities committed to the full discussion and exchange of ideas between and among its broad constituencies through meetings, publications, and its various electronic listservs and websites. The Organization's National Office is located in Champaign, Illinois and serves the association by planning conferences and meetings and publishing the various researches of its members. The main sources of support and revenues for the Organization are contributions, membership dues and subscriptions, and meeting income.

Subsequent Events

Subsequent events have been evaluated through November 13, 2024, which is the date the financial statements were available to be issued. There were no subsequent events requiring recognition or disclosure in the financial statements.

Accrual Basis of Reporting

The Organization has chosen to report on the accrual basis of accounting. Accordingly, revenue is recognized when earned and expenses are recognized when incurred in conformity with accounting principles generally accepted in the United States of America (GAAP). The financial statements are presented in accordance with Financial Accounting Standards Board (FASB) Accounting Standards Codification (ASC) 958 dated August 2016, and the provisions of the American Institute of Certified Public Accountants (AICPA) "Audit and Accounting Guide for Not-for-Profit Organizations" (the "Guide").

Use of Estimates in Preparing Financial Statements

The preparation of financial statements in conformity with GAAP requires management to make estimates and assumptions that affect certain reported amounts and disclosures. Accordingly, actual results could differ from those estimates.

Cash and Cash Equivalents

For purposes of reporting cash flows, cash and cash equivalents include all cash and highly liquid investments acquired with an original maturity date of three months or less. As of December 31, 2023 and 2022, $58,622 and $63,454 of cash and cash equivalents are designated by the board and are subject to board-imposed stipulations. As of December 31, 2023 and 2022, $37,184 and $34,904 of cash and cash equivalents are temporarily restricted and are subject to donor stipulations.

Investments

Investments consist of intermediate term bond funds, Bayer AG (BAYRY) stock, and money market funds, which are carried at the fair value of the underlying assets. Net appreciation (depreciation) in the fair value of investments, which consists of the realized gains or losses and the unrealized appreciation (depreciation) on those investments, is presented in the statement of activities in accordance with donor restrictions as investment income. Investment income is presented net of investment fees. The average cost method is primarily used to determine the basis for computing realized gains or losses.

Accounts Receivable

Accounts receivable are recorded primarily for outstanding invoices for membership dues and subscriptions. An allowance for doubtful accounts is based on an analysis of expected collection rates determined from experience. The Organization had an allowance for doubtful accounts of $0 and $0 at December 31, 2023 and 2022, respectively.

Inventory

The Organization's inventory of directories, research volumes, proceedings, and perspective magazines is carried at the lower of cost and market value. Cost is determined on the basis of first in – first out.

Property and Equipment

Property and equipment expenditures in excess of $2,500 are capitalized at cost. Donated property and equipment are capitalized at estimated cost or fair market value at the time of donation. Depreciation of the assets is computed using the straight-line method over their estimated useful lives. The range of estimated useful lives by type of asset is as follows:

Furniture and Equipment	3 - 7 years

Net Assets

Net assets of the Organization and changes therein are classified and reported as follows:

Net Assets without Donor Restrictions - Net assets that are not subject to donor-imposed restrictions and may be expended for any purpose in performing the primary objectives of the Organization. The Organization's board may designate assets without restrictions for specific operational purposes from time to time.

Net Assets with Donor Restrictions - Net assets subject to stipulations imposed by donors, and grantors. Some donor restrictions are temporary in nature; those restrictions will be met by actions of the Organization or by the passage of time. Other donor restrictions are perpetual in nature, where by the donor has stipulated the funds be maintained in perpetuity. As of December 31, 2023 and 2022, the Organization had no donor restrictions that were perpetual in nature.

Contributions

Unconditional contributions are recognized when pledged and recorded as net assets without donor restrictions or net assets with donor restrictions, depending on the existence and/or nature of any donor-imposed restrictions. Conditional promises to give are recognized when the conditions on which they depend are substantially met. Gifts of cash and other assets are reported with donor restricted support if they are received with donor stipulations that limit the use of the donated assets.

When a restriction expires, that is, when a stipulated time restriction ends or a purpose restriction is accomplished, net assets with donor restrictions are reclassified to net assets without donor restrictions and reported in the statement of activities as net assets released from restrictions. Donor-restricted contributions whose restrictions are met in the same reporting period are reported as net assets without donor restriction support. Contributions restricted for the acquisition of land, buildings, and equipment are reported as net assets without donor restriction upon acquisition of the assets and the assets are placed in service.

Contributed Services and Goods

Contributed services are reported as contribution revenue and as assets or expenses only if the services create or enhance a non-financial asset (for example property and equipment) or:

- Would typically need to be purchased by the Organization if the services had not been provided by contribution.
- Require specialized skills.
- Are provided by individuals with those skills (such as accounting, financial, construction, educational, electrical, legal, medical, and other services provided by accountants, investments advisers, contractors, teachers, electricians, lawyers, doctors, and other professional and craftspeople).

For the years ended December 31, 2023 and 2022, the value of contributed services meeting the defined requirements for recognition in the financial statements as outlined above were not material and have not been recorded on the financial statements. The Organization does receive free office space from the University of Illinois. However, the value for the use of this office space has been determined to be immaterial.

Income Tax Status

The Organization is a nonprofit association that is exempt from federal income tax under Section 501(c)(3) of the Internal Revenue Code. In addition, the Internal Revenue Service has determined that the Organization is not a private foundation as defined in Section 509(a)(1) and Section 170(b)(1)(A)(vi) of the Code.

The Organization has evaluated its exposure resulting from uncertain income tax position and determined the exposure is not material to the financial statements. In addition, the Organization is not aware of any tax position for which a significant change is reasonably possible within the next 12 months. Therefore, these financial statements do not include a liability for uncertain tax positions. Upon recognition of a liability for an uncertain tax position, the Organization would recognize interest expense and penalties in operating expenses.

The Organization files information tax returns in the U.S. federal jurisdiction and the state of Illinois. Its federal and Illinois information tax returns prior to calendar year 2020 are closed. The Organization does not have any tax returns currently under examination by either the Internal Revenue Service (IRS) or any U.S. state jurisdiction.

Membership Dues and Advance Subscriptions Collected

Membership dues and subscriptions are assessed and recognized as revenue based on the life of the dues or subscription.

Functional Allocation of Expenses

The costs of providing the various programs and other activities have been summarized on a functional basis in the statements of activities. The statements of functional expenses present the natural classification detail of expenses by function. Accordingly, certain costs have been allocated among the programs and supporting services benefited.

Expenses which are easily and directly associated with a particular program or supporting service are charged directly to that functional area. Compensation and payroll taxes & fringes have been allocated on the basis of estimates of time and effort.

NOTE 2 – ARRANGEMENTS WITH THE UNIVERSITY OF ILLINOIS

The Organization moved its offices to the University of Illinois at the end of 1999. Under an arrangement with the University, the employees of the Organization are employed by the University. The employees' pension and benefits are part of the University's plans. The Organization then reimburses the University monthly for the cost of its employees.

The University of Illinois holds some cash for the Organization. These "claim on cash" balances were $40,249 and $49,189 as of December 31, 2023 and 2022, respectively. These balances have been included with cash and cash equivalents.

NOTE 3 – CONCENTRATIONS OF CREDIT RISK

Financial instruments that potentially subject the Organization to credit risk consist principally of checking accounts, money markets accounts, and certificates of deposits at financial institutions. However, management continuously monitors the Organization's balances at financial institutions. The balances at each bank as of December 31, 2023 and 2022 were insured by the Federal Deposit Insurance Corporation (FDIC) up to $250,000. As of December 31, 2023 and 2022, $316,574 and $105,713, respectively, of the bank balance was not insured by FDIC. The "claim on cash" held with the University of Illinois as stated in Note 2 is not insured since it is not held in a financial institution. In addition, the Organization had undeposited funds of $324 and $600 at December 31, 2023 and 2022, respectively. Since these funds were not yet deposited with a financial institution it was not covered by FDIC.

NOTE 4 – INVESTMENTS

FASB ASC 820, *Fair Value Measurements*, establishes a framework for measuring fair value. That framework provides a fair value hierarchy that prioritizes the inputs to valuation techniques used to measure fair value. The hierarchy gives the highest priority to unadjusted quoted prices in active markets for identical assets or liabilities (Level 1 measurements) and the lowest priority to unobservable inputs (Level 3 measurements). The three levels of the fair value hierarchy under FASB ASC 820 are described below:

Level 1 – Inputs to the valuation methodology are based on unadjusted quoted prices for identical assets or liabilities in active markets that the Organization has the ability to access.

Level 2 – Inputs to the valuation methodology include quoted prices for similar assets or liabilities in active markets, quoted prices for identical or similar assets or liabilities in inactive markets, inputs other than quoted prices that are observable for the asset or liability, and inputs that are derived principally from or corroborated by observable market data by correlation or other means. If the asset or liability has a specified (contractual) term, the Level 2 input must be observable for substantially the full term of the asset or liability.

Level 3 – Inputs to the valuation methodology are unobservable and significant to the fair value measurement.

The asset's and liability's fair value measurement level within the fair value hierarchy is based on the lowest level of any input that is significant to the fair value measurement. Valuation techniques used need to maximize the use of observable inputs and minimize the use of unobservable inputs.

The following is a description of the valuation methodologies used for assets measured at fair value. There have been no changes in the methodologies used at December 31, 2023 and 2022.

Intermediate Term Bond Funds:	Valued at the net asset value (NAV) of shares held by the Organization at year-end based on readily determinable fair values, which are published daily and are the basis for current transactions.
Bayer AG (BAYRY) Stock:	Valued at the stated share price at year-end.
Money Market Funds:	Valued at cost plus accrued interest.

The preceding methods described may produce a fair value calculation that may not be indicative of net realizable value or reflective of future fair values. Furthermore, although the Organization believes its valuation methods are appropriate and consistent with other market participants, the use of different methodologies or assumptions to determine the fair value of certain financial instruments could result in different fair value measurements at the reporting date.

The following table sets forth by level, within the fair value hierarchy, the Organization's assets at fair value as of December 31, 2023 and 2022:

Assets at Fair Value as of December 31, 2023

	Level 1	Level 2	Level 3	Total
Without Donor Restrictions				
Bayer AG (BAYRY)	$ 7,384	-	-	$ 7,384
Money Market Funds	1,482	-	-	1,482
Designated by the Board				
Intermediate Term Bond Funds	26,797	-	-	26,797
Bayer AG (BAYRY)	29,536	-	-	29,536
Money Market Funds	5,926	-	-	5,926
Total	$ 71,125	-	-	$ 71,125

Assets at Fair Value as of December 31, 2022

	Level 1	Level 2	Level 3	Total
Without Donor Restrictions				
Bayer AG (BAYRY)	$ 10,288	-	-	$ 10,288
Money Market Funds	1,055	-	-	1,055
Designated by the Board				
Intermediate Term Bond Funds	25,263	-	-	25,263
Bayer AG (BAYRY)	41,152	-	-	41,152
Money Market Funds	4,220	-	-	4,220
Total	$ 81,978	-	-	$ 81,978

Investment income on the statements of activities for the years ended December 31, 2023 and 2022 are summarized as follows:

	2023	2022
Dividends	$ 3,783	$ 2,772
Capital Gains Distributions	-	7
Net Unrealized Gains (Losses)	(14,636)	(6,756)
Total Investment Income	$ (10,853)	$ (3,977)

This investment income, including unrealized gains and losses, are being reported on the statements of activities as investment income.

NOTE 5 – PROPERTY AND EQUIPMENT

As of December 31, 2023, Property and Equipment consists of:

Furniture and Equipment	$ 21,303
Less: Accumulated Depreciation	(20,202)
Property and Equipment, Net	$ 1,101

As of December 31, 2022, Property and Equipment consists of:

Furniture and Equipment	$ 21,303
Less: Accumulated Depreciation	(19,673)
Property and Equipment, Net	$ 1,630

Depreciation expense for the years ended December 31, 2023 and 2022 was $529 and $1,662, respectively.

NOTE 6 – NET ASSETS - BOARD DESIGNATED

Susan C. Eaton Scholar-Practitioner Memorial Fund

The Organization set up a memorial fund in honor of an author of a 1998 "Perspectives on Work" article who died on December 30, 2003. At its June 1, 2004 meeting, the Organization approved the establishment of an annual Susan C. Eaton Scholar-Practitioner award and grant to be paid to one or more qualified scholar researchers in even-numbered years or practitioners in odd-numbered years doing research in the labor and employment relations or related field. The Organization's Executive Board directed that 10% of any gifts received each year are undesignated and can be used for administrative expenses. As of December 31, 2023 and 2022, these designated funds totaled $12,779 and $16,779, respectively.

Kochan-Sleigh Best Dissertation Award Fund

On March 27, 2006, the Organization set up a designated fund with contributions from a member and matching contributions from General Electric. The fund is designated for a minimum of fifteen years and may be used to pay for the $1,000 annual best dissertation award and plaque. After the fifteen-year period, the fund may be continued or the amount remaining in the fund may be undesignated and become available to the Organization for unrestricted purposes. The Organization's Executive Board directed that 10% of any gifts received each year are undesignated and can be used for administrative expenses. As of December 31, 2023 and 2022, these designated funds totaled $34,200 and $35,319, respectively.

Jim Scoville Award Fund

On January 9, 2009, the Organization set up a designated fund with a $10,000 contribution from the University of Minnesota's Industrial Relations Center to honor a member and retiring professor. The fund was established to pay an annual award of $500 and a plaque for best paper on international and comparative employment issues. The Organization's Executive Board directed that 10% of any gifts received each year are undesignated and can be used for administrative expenses. As of December 31, 2023 and 2022, these designated funds totaled $818 and $2,176, respectively.

John T. Dunlop Public Policy Fund

On May 8, 2010, the Organization established a designated fund to subsidize a named John T. Dunlop Public Policy Session at future National Policy Forums and/or at the Organization annual meetings. The Organization's Executive Board directed that 10% of any gifts received each year are undesignated and can be used for administrative expenses. This fund will be so named for five or ten years, at which time the Organization's Executive Board will review the fund to determine whether to continue or modify the fund. As of December 31, 2023 and 2022, these designated funds totaled $18,127 and $16,503, respectively.

Gladys and Walter Gershenfeld Diversity & Inclusion Fund

On May 8, 2010, the Organization established a designated fund to honor the memory of long-time members and past presidents Walter and Gladys Gershenfeld for the purpose of supporting the Organization's electronic and print publications. The fund will be so named for ten years, at which time the Organization will review the fund's purpose and uses, and whether to continue or modify the fund. The Organization's Executive Board directed that 10% of any gifts received each year are undesignated and can be used for administrative expenses. As of December 31, 2023 and 2022, these designated funds totaled $19,494 and $17,939, respectively.

Quasi-Endowment Fund

During 2019, the Organization established a quasi-endowment fund for the purpose of supporting the Organization's general operations. The Organization's Executive Board directed that 20% of any gifts received each year are undesignated and can be used for administrative expenses. Funds designated by the Board of Directors to function as an endowment are voluntary and may be reversed by the Board of Directors at any time. Accordingly, they are reported as part of the net assets without donor restrictions. The Organization's Board designated endowment net assets exist as a permanent investment pool to enhance and sustain the operations of the Organization so that it can continue to be involved in the full discussion and exchange of ideas between and among its broad constituencies through meetings, publications, and its various electronic listservs and websites. As of December 31, 2023 and 2022, these designated funds totaled $35,462 and $45,373, respectively.

NOTE 7 – NET ASSETS WITH DONOR RESTRICTIONS

Net assets with donor restrictions at December 31, 2023 and 2022 are restricted for the following purposes or period:

	2023	2022
Subject to expenditures for specified purpose:		
Administer the Business of the Deans, Chairs, and Directors Committee (DCDC)	$ 8,910	$ 8,460
Administer the PhD Student Consortium	28,274	26,444
Total Net Assets with Donor Restrictions	$ 37,184	$ 34,904

Net assets were released from donor restrictions by incurring expenses satisfying the restricted purposes specified by the donors for the years ended December 31, 2023 and 2022:

	2023	2022
Administer the Business of the Deans, Chairs, and Directors Committee (DCDC)	$ -	$ 410
	$ -	$ 410

NOTE 8 – ENDOWMENT

Board designated net assets at December 31, 2023 and 2022 consist of an endowment fund established in 2019 for the purpose of supporting operations of the Organization. The Organization has adopted the provisions of FASB ASC 958-205-45, *Not-for-Profit Entities*, including *Net Asset Classification of Funds Subject to an Enacted Version of the Uniform Prudent Management of Institutional Funds Act, and Enhanced Disclosures for All Endowment Fund.* ASC 958 provides guidance on the net asset classification of donor restricted endowment funds for a non-for-profit organization that is subject to an enacted version of the State Prudent Management of Institutional Funds Act (SPMIFA) and also requires disclosures about endowment funds, both donor restricted and board designated endowment funds.

Interpretation of Relevant Law

The Organization has interpreted SPMIFA as requiring the preservation of the fair value of the original gift as of the gift date of the donor-restricted endowment funds absent explicit donor stipulations to the contrary. As a result of this interpretation, the Organization classifies as permanently restricted net assets (a) the original value of gifts donated to the permanent endowment, (b) the original value of subsequent gifts to the permanent endowment, and (c) accumulations to the permanent endowment made in accordance with the direction of the applicable donor gift instrument at the time the accumulation is added to the fund. The remaining portion of the donor-restricted endowment fund that is not classified as permanently restricted net assets is classified as net assets with donor restrictions - restricted for specified purpose until those amounts are appropriated for expenditures by the Organization in a manner consistent with the standard of prudence prescribed by SPMIFA.

Additionally, in accordance with SPMIFA, the Organization considers the following factors in making a determination to appropriate or accumulate donor-restricted endowment funds:

(1) The duration and preservation of the fund
(2) The purpose of the Organization and the donor-restricted endowment fund
(3) General economic conditions
(4) The possible effect of inflation or deflation
(5) The expected total return from income and the appreciation of investments
(6) Other resources of the Organization
(7) The investment policies of the Organization.

Endowment Net Asset Composition and Changes in Net Assets
The composition of endowment net assets and the changes in endowment net assets as of December 31, 2023 and 2022 are as follows:

	Without Donor Restrictions	
	2023	2022
Endowment Net Assets, Beginning of Year	$ 45,373	$ 45,462
Member Contributions	-	-
Investment Income	(9,911)	(89)
Endowment Net Assets, End of Year	$ 35,462	$ 45,373

Return Objectives and Risk Parameters
The primary objectives in the investment management for the portfolio assets is to provide ongoing stability and income for operations, preserving gift assets without subjecting them to substantial risk, and providing additional real growth through new gifts. The mix of investments in the portfolio is examined periodically.

Spending Policy
Spending from the fund should not exceed two to five percent annually depending on growth, general economic conditions, and organizational needs. The spending policy should be reviewed every five to ten years.

NOTE 9 – LIQUIDITY AND AVAILABILITY OF FINANCIAL ASSETS

The following reflects the Organization's financial assets as of the statements of financial position date, reduced by amounts not available for general use because of contractual or donor-imposed restrictions within one year of the statement of financial position date. Amounts available include donor restricted amounts that are available for general expenditure in the following year.

	2023	2022
Fiscal Assets at Year-End		
Cash and Cash Equivalents	$ 643,422	$ 435,876
Investments	71,125	81,978
Accounts Receivable, Net	12,609	1,773
Less Contractual or Donor-Imposed Restrictions		
Board Designated Funds	(120,880)	(134,089)
Donor Restrictions for Specific Purposes	(37,184)	(34,904)
Financial Assets Available to Meet Cash Needs for General Expenditure Within One Year	$ 569,092	$ 350,634

As part of the Organization's liquidity management, it has a policy to structure its financial assets to be available as its general expenditures, liabilities, and other obligations come due. Excess cash is generally held in checking accounts, money markets accounts, and certificates of deposits until it is required for operational use.

NOTE 10 – COMMITMENTS

The Organization signed a contract with the MotorCity Casino Hotel (the Hotel) in Detroit on November 7, 2019 to hold the 73rd Annual Meeting from June 1st through the 7th, 2021. Due to COVID-19, the Annual Meeting went virtual in 2020 and stayed virtual for 2021 and 2022. As a result, on September 10, 2020, the Organization did an addendum to the contract with the Hotel in Detroit to change the Annual Meeting date to May 31st through June 6th, 2022. On February 15, 2022, the Organization did another addendum to the contract with the Hotel in Detroit to change the Annual Meeting date to 2023, with a cut-off date of April 30, 2023. All other conditions of the contract including concessions and meeting space apply as originally negotiated. The Organization signed a contract with the New York Hilton Midtown in New York on December 21, 2022 to hold the 76th Annual Meeting from June 26th through July 1st, 2024. If this contract was cancelled at December 31, 2023, the Organization would have owed $295,927 to the New York Hilton Midtown in New York. If these contracts were cancelled at December 31, 2022, the Organization would have owed $121,311 to the Hotel in Detroit, and $184,954 to the New York Hilton Midtown in New York for a total of $306,265.

NOTE 11 – PENSION PLAN AND RETIREMENT CONTRIBUTIONS

The University of Illinois (the University) contributes to the State Universities Retirement System of Illinois (SURS), a cost-sharing multiple-employer defined benefit plan with a special funding situation whereby the State of Illinois (the State) makes substantially all actuarially determined required contributions on behalf of the participating employers. SURS was established July 21, 1941 to provide retirement annuities and other benefits for staff members and employees of State universities, certain affiliated organizations, and certain other State educational and scientific agencies and for survivors, dependents, and other beneficiaries of such employees. SURS is considered a component unit of the State of Illinois' financial reporting entity and is included in the State's Annual Comprehensive Financial Report as a pension trust fund. SURS is governed by Chapter 40, Act 5. Article 15 of the *Illinois Compiled Statutes*. SURS issues a publicly available financial report that includes financial statements and required supplementary information. That report may be obtained by accessing the website at www.surs.org.

Benefits Provided. A traditional benefit plan was established in 1941. Public Act 90-0448, enacted effective January 1, 1998, established an alternative defined benefit program known as the portable benefit package. The traditional and portable plan Tier 1 refers to members that began participation prior to January 1, 2011. Public Act 96-0889 revised the traditional and portable benefit plans for members who begin participation on or after January 1, 2011, and who do not have other eligible Illinois reciprocal system services. The revised plan is referred to as Tier 2. New employees are allowed six months after their date of hire to make an irrevocable election whether to participate in either the traditional or portable benefit plans. A summary of the benefit provisions can be found in the SURS' Annual Comprehensive Financial Report-Notes to the Financial Statements.

Contributions. The State of Illinois is primarily responsible for funding the SURS on behalf of the individual employers at an actuarially determined amount. Public Act 88-0593 provides a statutory funding plan consisting of two parts: (i) a ramp-up period from 1996 to 2010 and (ii) a period of contributions equal to a level percentage of the payroll of active members within SURS to reach 90 percent of the total Actuarial Accrued Liability by the end of fiscal year 2045. Employer contributions from "trust, federal, and other funds" are provided under Section 15-155 (b) of the Illinois Pension Code and require employers to pay contributions which are sufficient to cover the accruing normal costs on behalf of applicable employees. The employer normal cost was 12.32 percent of employee payroll during the period of July 1, 2021 to June 30, 2022. The employer normal cost was 12.83 percent of employee payroll during the period of July 1, 2022 to June 30, 2023. The employer normal cost was 12.53 percent of employee payroll during the period of July 1, 2023 to June 30, 2024. The normal cost is equal to the value of current year's pension benefit and does not include any allocation for the past unfunded liability or interest on the unfunded liability. Plan members are required to contribute 8.0 percent of their annual covered salary. The contribution requirements of plan members and employers are established and may be amended by the Illinois General Assembly. During the years ended December 31, 2023 and 2022, the Organization reimbursed the University of Illinois $23,182 and $22,349, respectively, for the employer's payments to SURS for eligible employees.

In addition to providing pension benefits, the State of Illinois provides certain health, dental and life insurance benefits to annuitants. This includes annuitants of the Organization. Substantially all State employees, including the Organization's employees, may become eligible for postemployment benefits if they eventually become annuitants. Health and dental benefits include basic benefits for annuitants under the State's self-insurance plan and insurance contracts currently in force. Life insurance benefits for annuitants under age 60 are equal to their annual salary at the time of retirement; life insurance benefits for annuitants age 60 or older are limited to $5,000 per annuitant. Currently, the State does not segregate payments made to annuitants from those made to current employees for health, dental, or life insurance benefits. These costs are funded by the State and are not an obligation of the Organization.

Employees of the Organization may also elect to participate in several tax deferred annuity plans and defined contribution plans. These are single employer plans under which benefits are provided to participating employees through contracts issued to each individual. Participation and the level of employee contributions are voluntary. The Organization is not required to make contributions.

LERA gratefully acknowledges the continuing support of its
2023–24 Major Sponsors and Organizational Members

Major Sustaining Sponsors

Illinois Climate Jobs Institute
International Union, UAW
Kaiser Permanente and the Alliance of Health Care Unions
UAW-Ford Joint Trust

Major University Sponsors

Cornell University, School of Industrial and Labor Relations
Florida International University, College of Business
Massachusetts Institute of Technology, Institute for Work and Employment Research
Michigan State University, School of Human Resources and Labor Relations
Pennsylvania State University, School of Labor and Employment Relations
Princeton University, Industrial Relations Section
Rutgers University, School of Management and Labor Relations
University of Illinois Urbana-Champaign, School of Labor & Employment Relations
University of Minnesota, Center for Human Resources and Labor Studies

Organizational Members

American Federation of Government Employees, AFL-CIO
American Federation of Teachers, AFL-CIO
American Train Dispatchers Association
Boston University, Human Resources Policy Institute
Castrey Dispute Resolution
Centre for Employment and Labour Relations Law / Melbourne Law School
Cornell University, Scheinman Institute on Conflict Resolution
Cornell University, School of Industrial and Labor Relations
Federal Mediation and Conciliation Service
Florida International University, College of Business
International Association of Machinists and Aerospace Workers
International Brotherhood of Electrical Workers, Local 11
International Union, UAW
JAMS
Jobs With Justice
Kaiser Permanente and the Alliance of Health Care Unions
Keystone Research Center
Massachusetts Institute of Technology, Institute for Work and Employment Research
Michigan State University, School of Human Resources and Labor Relations
New Jersey School Boards Association
Pennsylvania State University, School of Labor and Employment Relations
Portland General Electric
Portland State University
Princeton University, Industrial Relations Section

Rhode Island Council 94, AFSCME, AFL-CIO
Rutgers University, School of Management and Labor Relations
Taiwan Institute of Labor, Occupational Safety and Health
UAW-Ford JTC
United Mine Workers of America
United Nurses Associations of California/Union of Health Care Professionals
U.S. Department of Agriculture
University of California, Los Angeles, Anderson School of Management
University of California-Berkeley, Center for Labor Research and Education
University of California-Berkeley, Institute for Research on Labor and Employment
University of Illinois Urbana-Champaign, School of Labor and Employment Relations
University of Minnesota, Center for Human Resources and Labor Studies
University of Toronto, Centre for Industrial Relations and Human Resources
University of Wisconsin School for Workers
Utility Workers Union of America, Local 470-1
Virgin Islands Labor Management Committee
Wayne State University, Labor@Wayne
W.E. Upjohn Institute for Employment Research

For more information about organizational membership with LERA, please visit the LERA website: https://lera.memberclicks.net/organizational-membership.

ALPHABETICAL LIST OF AUTHORS